Eugene Allen Smith's Alabama

Eugene A Smith

EUGENE ALLEN SMITH'S ALABAMA

How a Geologist Shaped a State

AILEEN KILGORE HENDERSON

FOREWORD BY LEWIS DEAN

NEWSOUTH BOOKS
Montgomery | Louisville

ALSO BY AILEEN KILGORE HENDERSON

The Summer of the Bonepile Monster (1995)

The Monkey Thief (1997)

Treasure of Panther Peak (1999)

Stateside Soldier: Life in the Women's Army Corps, 1944–1945 (2001)

Tenderfoot Teacher: Letter from the Big Bend 1953–1954 (2002)

Hard Times for Jake Smith (2004)

NEWSOUTH BOOKS
105 S. Court Street
Montgomery, AL 36104

 Published in the United States by NewSouth Books, a division of NewSouth, Inc., Montgomery, Alabama. All images in this book are reproduced here courtesy of the William Stanley Hoole Special Collections Library, University of Alabama, Tuscaloosa. Except as noted, all photographs were taken by Eugene Allen Smith.

LIBRARY OF CONGRESS CATALOGING-IN-PUBLICATION DATA
Henderson, Aileen Kilgore, 1921–
Eugene Allen Smith's Alabama : how a geologist shaped a state / Aileen Kilgore Henderson.
p. cm.
Includes bibliographical references and index.

ISBN-13: 978-1-58838-243-6
ISBN-10: 1-58838-243-5

1. Smith, Eugene Allen, 1841-1927. 2. Smith, Eugene Allen, 1841-1927—Influence.
3. Geologists—Alabama—Biography. 4. Alabama—Officials and employees—Biography.
5. College teachers—Alabama—Biography. 6. University of Alabama—Faculty—Biography.
7. Geology—Alabama—History. 8. Scientific expeditions—Alabama—History.
9. Industrialization—Alabama—History. 10. Social change—Alabama—History. I. Title.
QE22.S598H36 2011
551.092—dc23
[B]

2011030787

Design by Randall Williams

Printed in the United States of America by Sheridan Books.

In memory of my father,

William O. Kilgore Sr.,

who first told me about Eugene Allen Smith in the 1930s.

Eugene Allen Smith's famous Studebaker wagon, now displayed at Smith Hall, University of Alabama. The Alabama Geological Survey team used the wagon from 1883–1904 (see pages 54–55).

Contents

Eugene Allen Smith (1841–1927)

Foreword

Lewis Dean
Geological Survey of Alabama

Over the past decade, Aileen Kilgore Henderson has been a welcomed visitor to the library in the office of the Geological Survey of Alabama, on the campus of the University of Alabama. It would take an established writer, educator, and someone with an incurable optimism to research the records available on Eugene Allen Smith and produce an engaging narrative recording the life and career of one of Alabama's most distinguished public servants. This book, which came from Aileen's endeavor, shows she was well suited for the task.

Soon after his appointment as professor at the University of Alabama, Smith was placed in charge of implementing an 1873 legislative act to revive the state geological survey of Alabama. Smith's archived field notes, papers and published works, covering his tenure as state geologist, are a unique record of geological research and documentation of Alabama cultural history. The Smith collection contains original field notebooks, which in the 1930s were compiled into a seven-volume set of transcribed records of 2,300 pages by Walter B. Jones, Alabama state geologist and protégé of Smith. This collection documents efforts by Smith in numerous benchmark reports describing every geologic province of the state. During Smith's epic 54-year career as state geologist, from 1873 to 1927, he published more than 120 titles on Alabama and Southeastern geology.

Smith also recorded initial collections for paleontology, natural history, and extensive photography which became the nucleus of the present Alabama Museum of Natural History. One of the legacies of Smith's work is the recognition that his research is of the highest standards, and his collections and accompanying field notes are an extensive primary resource on a variety of topics regarding the geology and cultural history of Alabama.

Eugene Allen Smith's Alabama is not a conventional biography. Rather, it is a narrative account of the scientist's life, primarily in Smith's own words, revealing the core aspect of his professional career in undertaking a geological survey of the state. The value of the book, however, is how Henderson enables the reader to come to admire Smith and to recognize that he did, indeed, change Alabama. Henderson relies on Smith's own words, and even a representative sampling of his photographs, to let the reader know Smith and his work.

It can safely be said that Smith was remarkable: short in stature, but a giant of a man. He believed in progress. His life and his work testify to the conviction that society and individuals can build a better world. He had the essential characteristic of a scientist: curiosity. He had the physical and mental stamina to pursue knowledge wherever it led. He was also a teacher, not only of university students, but of secondary school students, the general public, the political classes, industrialists and entrepreneurs, and his academic colleagues. He pursued knowledge of Alabama's natural resource potential, believing that if disseminated and acted on, the knowledge could transform the state's economy and the lives of his fellow Alabamians. Henderson provides brief glimpses of an even more well-balanced man, in his love of music and competitive sport, and of his never-flagging support of his wife and children.

Like Smith, Henderson has done her state a service. *Eugene Allen Smith's Alabama* will reintroduce to the reading public a preeminent Alabamian who in his own time had a positive influence in shaping his native state and who left an enduring legacy of science and service.

Introduction

Scattered in a muddy field in northern Kentucky, the soldiers of Company K, 33rd Regiment, Alabama Volunteers, occupied themselves with camp chores or sprawled on the ground sleeping. After an eighteen-day forced march from Tupelo, Mississippi, the exhausted men expected a confrontation with the enemy at any time. Since their arrival they had heard the sounds of battle, and they had seen the fine-looking soldiers in blue who had been captured by Confederate forces. Company K waited on September 18, 1862, to be called to the battle.

Oblivious to his surroundings, a wiry, diminutive private named Eugene Allen Smith, age twenty-one, sat on a boulder writing to his physician father in Prattville, Alabama. With typical vigor, he wrote steadily, summing up the eventful weeks since his last letter: he had recovered from the fever afflicting him when the company left Mississippi, from the suffering on the long marches, sometimes twenty-five miles a day, from marching in drenching rain and sleeping in the mud without blankets, from washing his meager rations down with water out of mud holes not fit for hogs to wallow in, and from walking the soles off his only pair of shoes.

More interesting, however, than any of this news, at least to Private Smith, were the marvelous things he had seen along the way north—abundant fossils, rocks with perfect leaf and shell impressions, and quartz crystals of unusual beauty. While his comrades lightened their loads by discarding their most essential possessions, Eugene weighed down his knapsack with rocks and other choice specimens. He hoped to find a way to get them back home to add to the collection he had begun as a child. He lamented how difficult it had been passing so near Mammoth Cave and not catching even a glimpse of it.

Within the next month, Private Smith survived the bloody battle of Perryville and by the close of the year had been discharged from service in the Confederate army. An order from Richmond appointed him professor of

Jane (Jennie) Henry Meredith Garland (Mrs. Eugene Smith). Studio photograph by W. E. Armstrong of Nashville, Tennessee.

tactics at the University of Alabama, his alma mater. Until the end of the war, he remained within the bounds of Alabama training recruits. As Captain Smith, he saw the burning of the university when Federal troops captured Tuscaloosa, April 3-4, 1865, and he retreated with the cadets to Marion at the order of the university president, Dr. Landon Garland.

After the Confederate surrender less than a week later, Smith re-entered private life confident about his future. With an excellent educational background—private school in his hometown of Prattville followed by high school in Philadelphia, then a bachelor of science degree from the University of Alabama—he now turned his thoughts toward post-graduate education. Borrowing $1,400 from his grandmother's sister in Connecticut, he sailed for Europe to attend the great universities of Germany. Considering his interests and his later career, he probably concentrated on the subjects of chemistry, botany, geology, mineralogy and soils.

In 1868 he returned to the American South, as a skilled violinist, fluent in languages, and saturated with the culture of the Old World. Tucked in his trunk, in addition to a master of arts degree, was a doctorate with highest honors from Heidelberg University.

For three years he taught chemistry at the University of Mississippi, Oxford, and worked as assistant to Dr. Eugene W. Hilgard, state geologist of Mississippi, who remained his mentor and friend as long as Hilgard lived. Both of these appointments prepared Smith well for his life's work that began in 1871 in his native state and ended with his death in 1927, a span of more than five decades. What he accomplished, against monumental odds, became the catalyst that transformed Alabama from an aimless and poverty-stricken agricultural state to an industrial giant. How he accomplished what he did, with very little support and hardly any money, gave this five feet four geologist a stature of mythical proportions in the history of the university and the state. Jolting his way through every Alabama county in a mule-drawn wagon he came to know the state and its people more intimately than any other man. In his journal, his reports and hundreds of letters, he left a vivid picture of an Alabama in transition. Adding to the written record is his striking photographic documentation, beginning in 1885, of what he saw during his years of travel.

From Oxford, Smith took with him his dog Flirt, but left behind many friends and his long-time sweetheart, Jane Henry Meredith Garland (Jennie), daughter of Dr. Landon Garland, the wartime president of the University of Alabama and at the time a professor at the University of Mississippi. Because of his ability, education, and genial personality, Smith had other, more lucrative and culturally attractive opportunities than an appointment in Tuscaloosa. For one, a friend from his years in Europe, James Edward Graybill, pressed him to relocate to New York City where he could stay with Graybill and go into partnership with William H. Wahl, their mutual friend, also from their student days in Germany.

Instead Smith chose to come home to Alabama as professor of geology and mineralogy at his alma mater a mere six years after its destruction. In the one building that the school had managed to erect, he had an ill-equipped basement laboratory and a poorly lighted classroom. He wasted no time with regrets. He remembered with pleasure his years at the school and settled easily into the routine of teaching. He sent to Prattville for his geological collection, an atlas, and his books. He ordered more books, including Massey's *Shakespeare's Sonnets,* Lyell's *Principles of Geology*, and a life of Robert Burns, from an importer of books in New York. Before long he was climbing the bluffs above North River collecting specimens of Alabama snow wreath for Asa Gray, the well-known Harvard botanist.

Smith and two other professors undertook to replace the collection of the destroyed library that had been one of the finest in the South. He also filled the position of librarian and, for a time, added acting commandant of the school to his titles. In the coming years, he helped found the school's athletic program and organize the university symphony orchestra in which he played second violin.

In these early years not only did the school lack basic equipment and textbooks, but in 1872 faculty salaries also lagged four months behind schedule. Student enrollment fell short of expectation due to the low price of cotton and pervasive poverty. Smith, convinced that Alabama's recovery depended on the development of the state's natural resources, pushed the legislature to reactivate the state geologist position. His efforts succeeded in 1873 when the legislature designated him Alabama state geologist, allowing him $500 annually for travel and incidental expenses, and an initial appropriation of $3,000 for the purchase of a camp wagon plus equipment and laboratory chemicals.

Smith's eight-year courtship of Jennie finally culminated in marriage on July 10, 1872, in Oxford. Relocating to Tuscaloosa was a homecoming for Jennie—she had been less than a year old when the Garland family traveled by wagon from Virginia to Tuscaloosa where her father assumed the chair of English literature at the university and later the presidency. The newlyweds rented a place in town for a while, but were soon able to set up housekeeping on campus.

Alabama had been without a geologist since the death of the able Michael Tuomey in 1857. Smith had little to build on as most of Tuomey's scientific papers and his collections burned with the university. Other setbacks included the failure of the camp wagon and the expense account to materialize. Smith had to arrange a means of travel himself and pay expenses out of his own funds. He requested passes from railway officials that saved money but limited the range of his investigations. He had to maintain any equipage he rented. He had the wagon's axles greased every morning and paid for blacksmith services (twenty cents to shoe one mule foot and thirty-five cents for mending the wagon's coupling pole) and fifty cents a day for a driver who also cooked and took care of the mules. He also paid varying sums to be ferried across the many bridgeless streams (night crossings cost more). Besides these adversities, Smith coped with the extremes of summer weather, the inevitable chills and fevers, and the plagues of seed ticks and chiggers that kept sleep at bay.

In the almost six decades from Dr. Smith's appointment as a University of Alabama professor until his death is 1927, the "Little Doc" proved himself to be a giant. He was instrumental in transforming Alabama from its impoverished post-Civil War agricultural economy into the state with the most diverse economy in the South. In addition to his duties as professor of geology, mineralogy, chemistry, and agriculture, Dr. Smith enriched the life of the University of Alabama: he organized its symphony, rebuilt its library, and helped found its athletic program. He organized the annual field day and supported football and baseball. He personally watched university buildings being willfully destroyed by a Union general in April 1865. (His influence resulted in almost fifty thousand federal acres been set aside for reparations for burning of the university.) Smith rebuilt the core collection of what became the Alabama Museum of Natural History in Tuscaloosa; the antebellum collection had also been destroyed by Union troops. Smith's many contributions to the museum culminated in the dedication of the current museum building in 1910, a building that he designed and which bears his name. Thanks in large part to Dr. Smith, the museum ranks among the finest natural history museums in the country. He colloborated in publishing the first complete listing of Alabama plants, and he was the first to recognize the archeological significance of Moundville.

It was, however, his service with the Geologic Survey of Alabama and as state geologist that enabled him to amass an unparalled storehouse of knowledge that provided opportunities for Alabama and its people. He advocated a mixed economy; and he identified resources to realize economic diversity—especially mineral resources, but also the potential for linking Birmingham and Tuscaloosa by a lock and dam system. He not only disseminated this information in scientific papers and official reports to the state government, but also in hundreds of articles in the popular press. Consequently, those seeking investment opportunities, especially Northern industrialists, built Alabama's industrial sector on coal, iron, graphite, mica, kaolin, cement, soapstone, timber, fertilizer, and water power. In 1892, he produced the first accurate state map; this resulted in his appointment to the five-man Alabama Highway Commission in 1911. His map remained the standard until 1926.

The bulk of this book follows Smith's annual expeditions for five decades. Smith's words—and his photographs—reveal the man's physical stamina, intellectual curiosity, and scientific discernment. Most importantly, they reveal the optimism of the Progressive Era and its conviction that we can build a better tomorrow. Eugene Allen Smith embodied that spirit, and he changed Alabama.

Talladega Creek near Dr. Taylor's mill fell "in several beautiful cascades ten feet and more in height," Smith wrote in 1875. He made this photograph on March 18, 1908 (see page 16).

Eugene Allen Smith's Alabama

Smith first visited the Cahaba River below Pratt's Ferry, Bibb County, in 1877. He made this photograph on August 12, 1889.

1

The First Decade, 1873–83

1873

Jefferson and Shelby Counties*

Early on, Smith focused his attention on the state's crippled iron industry. Federal troops had destroyed fifteen of Alabama's sixteen producing furnaces along with their forges, shops and foundries.

His first trip as state geologist included a tour of Henry DeBardeleben's ore works in the fledgling Birmingham area. Smith noted how the limestone, needed as a flux for smelting the iron ore, conveniently underlay the ore itself; how DeBardeleben, destined to become one of Alabama's most powerful industrialists, leased convicts from the state to work the ore; and how the furnaces were not as productive as had been expected.

Smith also took a brief tour of the coal beds during which he evaluated the wide range in quality of the coal. Along his route he met two men who typified the hundreds of dreamers who crossed his path in later years. A Mr. Freeman invested his money and his time sinking shafts on the side of a ridge already pitted by ancient excavations. He confided to Smith that early Americans had made the excavations searching for something valuable, and he determined to find it whatever it was. From studying the sections in one of the deep shafts, Smith knew that the earth concealed nothing worth digging for. Farther along the way a Mr. Cawlfield proudly showed off his volcanic craters that Smith identified as lime sinks. Despite the geologist's advice, both men remained fixed in their respective opinions.

When he summed up this beginning summer's work in his report to Governor David P. Lewis, the first of the many governors he dealt with, Smith pointed out how handicapped he had been by the complete lack of funds.

*The text follows Smith—through journals, letters, diaries, logs, receipts, and other documents—on a year-by-year basis. Following each date subdivision are listed the counties or other localities mentioned in the text that follows. A Gazetteer at the end of the book will help contemporary readers with the place names Smith mentions. Some of the smaller towns no longer exist.

He expressed the hope that the money would be more forthcoming in 1874. However, the shortcomings of the summer brought out Smith's ability to look on the bright side of events. If his train crept along fifteen hours late, he exulted in the opportunity to botanize from the car window. If he drove through the geologically dull piney woods in August, he studied the flowers in bloom everywhere "in great abundance and beauty, as only the piney woods flowers can be."

At the close of 1873 the first of Eugene and Jennie's four sons was born. He was christened Eugene Allen Smith Jr.

1874

CALHOUN, CHAMBERS, CHILTON, CLAY, CLEBURNE, COOSA, ELMORE, LEE, MACON, RANDOLPH, TALLADEGA, AND TALLAPOOSA COUNTIES

Smith began the 1874 geological tour in mid-August from Prattville. Soon after starting out, he visited with a Mr. Gibbons who showed him a gully so sterile that only *Kalmia latifolia* (mountain laurel) grew at its edges. Smith was interested to see that Gibbons had fertilized his flourishing crops with pine straw, a system advocated by Dr. Hilgard of Oxford, a noted soil scientist.

Several aspects of the Blue Creek area tempted Smith to linger. Dr. Dawson guided him to an Indian fort with rock walls so ancient that trees two feet in diameter grew on them. Near Verbena Station, Mr. Spencer panned the creek for gold using a hollowed-out poplar log for a rocker. He showed Smith how washing a shovelful of dirt left several tablespoons of heavy black sand. Then, carefully washing the black sand, he salvaged the gold. In a relatively short time, he had accumulated about eighty dollars in large and small lumps.

The numerous caves under the creek's rock ledges had provided hiding places for Confederate deserters, Smith remembered. At the curious "Turn Around" near Verbena, a ridge of micaceous schist only fifty yards across split the creek in half for about a mile, then the two streams turned around and rejoined. The excited locals mistook the iridescent tarnish of this schist, which with quartz was the prevailing rock, for the "blossom" of copper.

On to Clanton, where Smith inspected two pits that produced large sheets of mica, and collected specimens of the beautiful sandstone that came in two forms, the pure white hard quartzite, and the friable "perfectly white sharp sand" used locally for scouring.

When the geological wagon and tent set up at the Shelby Iron Works, visitors of all kinds besieged it. Smith wrote: "Negros and whites, old and young, crowded about the tent, wanting to buy wares, to have pictures taken, to see the show. In fact the population seemed well nigh demented." Smith was cordially received at the Shelby Works. As an employee conducted him on a tour, he noted 2,400 tons of pig metal piled about the yard awaiting a more favorable market. A Mr. Preble discoursed on the raw materials the works used, the peculiarities of the iron each produced, and presented Smith with examples of their finished products. The assistant superintendent showed off their new furnace stressing that it had been "constructed by the most approved plan." Toward the close of the day, Smith took tea and had a pleasant chat with Mr. Willoughby, a company official and a "thorough Yankee." The host asked Smith to take samples of his well water for testing; it had such an offensive odor he was sure that it contained "something."

At the Gant marble quarry, which had formerly produced a very fine marble, Smith found the buildings in ruins and the quarry filled with water due to neglect caused by litigation after Dr. Gant's death. In a nearby branch, Smith saw masses of black limestone shot through with white veins; his guide, Dr. Hill, had cut and polished a slab of it into a handsome tabletop. In this same area he visited slate quarries opened during the war by the Welshman, Elliott, to furnish the Confederate States arsenal at Selma. This last day of August was unsettling, however, as Smith wrote in his journal that night: "I have been troubled today to account for the positions of the various rocks which I have met." He sketched how he thought the rocks should lie, but he had no proof that he was right.

From Dr. Hill's, he traveled over a superior road through fertile land and beautiful scenery to Talladega County, disposing of Sylacauga along the way as "a dilapidated old town on top of a red clay hill." In Talladega, Smith's Uncle James joined the expedition to explore the county that Smith termed "a garden spot." Dr. Taylor offered to guide them to a gold mine that had been working for about ten years, but Uncle James had a headache and Smith returned him to his home instead.

Near Oxford, Smith toured the Woodstock Iron Works where he did not feel as welcome as he had at the Shelby furnace. He collected specimens and inspected with interest a large mass of pure iron that he identified as a possible meteorite. Returning to town ("much relieved") he spent the rest of the day testing specimens of gold, silver, and copper ores brought in by the local people. As usual nothing of value appeared, and as usual his analyses left everyone unhappy.

On the way to Boiling Spring, Smith made note of the many clear springs gushing from the hillsides—"a great treat for the thirsty traveler." As the wagon bumped along, he was so absorbed in classifying the rocks he had collected he missed the turn-off to Boiling Spring and had to change his plans. He continued to the gold valley surrounding Arbacoochee, staying a night with Mr. Douthard, "a jolly old bachelor" who had lived there since the days of the gold rush. Below the Douthard house, the valley was scarred with countless pits left from the 1830–40 boom time when the town swelled to a population of five thousand occupying a hundred houses, many tents and shacks and patronizing twenty stores, five bars, two hotels, a race track, a school, and two churches. Now only two mines were functioning on a small scale.

Before leaving Cleburne County, Smith visited a copper mine that had been abandoned since the beginning of the war. He found such plentiful evidence of copper scattered on the ground that he judged a little work and money invested would return a worthwhile profit. He went by Wedowee, to pick up his mail but had none. Perhaps his disappointment colored his description of Wedowee as situated on a high ridge with "very little attractive about it being an old broken down town galvanized with some show of life" because former Governor W. H. Smith lived there and kept a store.

September 11, 1874, the former governor, now a circuit judge, gave Smith a guided tour of his Pinetucky gold mine a few miles outside Arbacoochee. Most of the area's surface had been mined and then tunneled under. In his operation, Judge Smith used convicts leased from the state at twenty-five cents a head per day. Like DeBardeleben's iron ore works that also used convict labor, Judge Smith's gold mine was not successful. Smith suggested in his journal that this was due partly to an unreliable labor force and partly because Judge Smith was an absentee owner. Other reasons might be found in the character of the man. One historian described him as "narrow, vindictive, and . . . unfit for office." During his governorship (July 1868–November 1870), Alabama's public debt tripled as a result of his dubious financial dealings that led eventually to the state's bankruptcy. When he was defeated in his bid for reelection, with the support of the occupying Federal troops he barricaded himself in the governor's office for several weeks, refusing to concede. Now he had retired to Randolph County and was dabbling in gold mining. Smith, diplomat that he was, took no notice of these past events in his journal.

That night Smith camped at ghostly Chulafinnee, a town that owed its existence to the discovery of gold in 1835. Smaller than Arbacoochee but more imposing, its brick buildings failed to guarantee its permanence. Legend had it that the King brothers, later of Texas ranch fame, were among the gold hunters in Chulafinnee. One celebrated his leave-taking to fight for the Confederacy by allegedly dynamiting the gold field. Smith hoped to see the exact spot where gold was first discovered, but he could find no one to show him.

Near Louina, Smith noted the clay bed of porcelain-ware quality that had been known since Michael Tuomey's day but had never been used. At Flatrock on the Tallapoosa River, he examined the extensive spread of granite that gave the village its name and provided a flourishing trade in millstones. On the road out of the valley of Chulafinnee, he found "thousands on thousands of little garnets" covering the ground like "bog ore bullets."

The great chunk of pure iron at the Woodstock furnace had been found in this area. Smith had come around to thinking that unique mass of iron might be the bloom (wrought iron ready for working) accidentally dropped from one of the nearby forges in olden days. He heard from the locals that after the forges shut down, the people of the countryside carted off many loads of bloom to use for their hardware needs. Yet he did not yield the possibility that it was of meteoric origin.

On September 14, Mr. Wethers showed Smith an ancient pit where large plates of mica strewed the ground. Wethers confided that he had dug there until he uncovered "a green sandy substance which . . . galvanized his spade." What the man meant by his spade being "galvanized" baffled Smith, but he gladly accepted a large tourmaline crystal that Wethers had uncovered near his mica dig.

Smith wrote at length about Wood's Copper Mine, a flourishing operation in Cleburne County. Five or six shafts, twenty-five feet deep and twenty feet apart, were sunk into the lode. Windlasses and buckets brought the ore up;

it was piled on a platform where the best of it was bagged for hauling to Carrollton, Georgia, forty miles away, then shipped to Baltimore. The expense of bringing a twenty-five hundred pound ton to market was thirty-five to forty dollars; the selling price was sixty dollars. In four months the mine had produced three hundred tons. Mr. Wood told Smith that he planned to increase his production by marketing a concentrate of the poor-grade ores that at the present time were cast off.

Randolph, Chambers, Tallapoosa, and Chilton counties included a number of so-called de Soto mines supposedly dating from the Spanish expedition of 1540. One of these, a mica excavation that Smith examined, measured fifty feet across and fifteen feet deep. Large pines grew out of the debris piled around it. He marveled that men could not realize the impossibility of Hernando de Soto and his armored followers developing the many different mines claimed for them over such a widespread territory.

All the creeks originating in Clay County flowed into either the Coosa or the Tallapoosa rivers including Talladega Creek that Smith time and again lauded for its beauty. He wrote, "It rises in the highlands east of the Blue Mountain ridge, flows down the valley southwest for ten or twelve miles, then turns northwest and cutting through the quartzites, argillites and other rocks of the western border flows through the Silurian valley of Talladega towards the Coosa not far from Childersburg. Where this creek cuts through the metamorphic ridges, several beautiful cascades are found."

Smith savored unusual names such as the Ketchapedrakee that cut a gap through a ridge near an eighty-foot deep shaft sunk through mica schist by the Tennessee Mining Company in an unsuccessful search for copper. A fifty- to sixty-foot deep tunnel burrowing into the same ridge had pleasant tasting springs trickling out of the rock crevices. The healthful water and the rugged, beautiful scenery, Smith suggested, could make the spot into a popular resort if it were more accessible. Candutchkee, an old Indian settlement near Ashland drew his attention as did the bountiful springs at the foot of Blue Mountain and the handsome tourmaline crystals and garnets up to two inches in diameter that he found along the way.

The Socopatoy area in Coosa County showed several outcroppings of a true granite that made a handsome, fine quality building stone that Smith praised for the ease with which the enormous blocks could be cut and quarried. He noted that the Coosa County gold deposits showed little activity even though persistent panning in the small streams could still turn a profit.

At Munford, Smith rode six miles into the mountains to see Mr. Stewart's silver mine that had been widely publicized. It turned out to be a twenty-foot deep hole with no sign of silver anywhere. A story familiar to Smith had captivated Stewart: mysterious markings on a beech tree pointed to the opening of a rich silver lode mined by Indians. They captured a black man, blindfolded him, and took him to the marked tree where they removed the blindfold and forced him to load the silver. Then they blindfolded him again and made him haul the silver down into the valley. When he finished, the Indians punched out his eyes and left. How did Stewart know this? Someone conveniently found an elderly, blind black man living in the area who supposedly told all, assuring everyone that the mysterious signs marked the entrance to the silver mine. Excavations dug by believers pockmarked the surrounding hills. Smith's advice was "let the thing alone."

Smith rode past the burial place of the Indian chief Chennibee on his way to visit Mr. Seay, who claimed to be on the track of a silver mine of untold wealth. He too had pieced together hints dropped by an elderly Indian friend forty years earlier, and though he confessed to Smith that he found no evidence to support the story, he continued to dig. This time Smith kept his doubts unspoken except to his journal that night.

At the end of September 1874, after traversing hollows along Hillabee Creek "full of rolled pebbles of quartz the size of one's fist," the geological wagon established camp at Mr. Pinkston's. Smith rode to Dudleyville for groceries (he found none) and to pick up his mail (he had three letters). Later his host showed off his collection of rocks and minerals, catching Smith's eye with a fifty-two-pound crystal of corundum, an extremely hard mineral closely related to rubies and sapphires. Smith recalled that the Reverend Mr. C. D. Smith of Franklin, North Carolina, in 1871 discovered corundum two miles west of Dudleyville. The Reverend Smith had written Dr. Smith about his discovery: "I am persuaded that judicious explorations will bring to light gems of great value." Smith managed to acquire Pinkston's specimen of corundum for the university.

In October near Mill Town, Smith was impressed with William Jackson's mill and quarry. Jackson made gravestones and monuments—a large collection cluttered his yard—but Smith felt that the rocks he used took such a beautiful polish so easily that they could be made into mantels and tabletops that would rival those made of the more expensive marble. He wrote of a "hornblendic schist stuck full of large garnets like plums in a pudding, then greenish soapstone approaching serpentine, then chlorite slate. Some of the serpentine . . . passes into a light green rock full of radiating crystals of actinolite. [It] . . . is quite soft and easily worked except that full of garnets. Mr. Jackson says the garnets are as hard as flint and wear out his saws directly but the mottled appearance, dark red upon green is attractive to some purchasers and sells for a better price than the more beautiful green actinolite rock . . . In the softest part of the rock are traces of the old Indian potters' work, in fragments of broken utensils, and marks of the cutting instrument in the rock itself."

Around Talladega, with Mr. Watts as guide, Smith explored the ruins of the Montgomery Mining Company. The roasting and smelting works, the buildings, the excavations, all told "of greatness gone by." Pyrite still lay in piles about the opening to the eighty-foot shaft now partially filled with water.

Near the deserted village of Oak Bowery, which Smith thought must have once been the "loveliest of the plains," Mr. Billingslea turned out blocks of high-grade soapstone to be used in lining kilns for lime burning. On the way to LaFayette, he noted the flourishing farms, "a sight . . . rarely seen in Alabama since the war." When his rig rolled into town it caused a sensation. Crowds of the curious bombarded him with questions, and at the hotel he was greeted with "What are you peddling?"

In this area noted for its quartz crystals, Smith had a number of people he wanted to see, but they were all gone to a camp meeting four miles south of Wedowee. On arriving there, he "found an immense crowd assembled, People were bivouacking every where in the bushes and in temporary sheds." When the geology rig hove into view, the crowd left the daguerreotype artist's tent to swarm around Smith to see what he had to offer. Realizing he would have no peace camping there, he continued to the nearest house, Mr. Bonner's, where he sought permission to set up in the grove.

Finding no one home, Smith returned to the meeting to seek Bonner. Instead he met a mutual acquaintance who told him to go ahead and camp; he would clear the matter with Bonner. "I had just put up my tent," Smith wrote later, "when Mrs. Bonner and a crowd of ladies came in, the former in high dudgeon and sniffing the air and muttering . . . about the impropriety of putting up a tent before one's house without so much as asking permission, etc. I heard enough to learn the drift of her remarks, and going up to her endeavored to explain the situation: but she turned up her nose and walked into the house without saying a word. I then pulled down the tent, and marched back to the woods out of sight of the grounds."

The crowds found him, however. They raised a hubbub exceeding the one at the Shelby Iron Works, demanding to have their pictures taken, clamoring to buy whatever he was selling. Later in the night, Smith's sleep was disrupted by men sneaking around the camp looking for whiskey, tobacco, or cards. Early next morning, he packed up and escaped, but not before he had several promises of specimens to be sent to him at the university. Mr. Osborne, who claimed he had panned pieces of lead soft enough to cut with a knife from a branch near his house, was among the promisers. Mr. Finch was not. The specimen of chalcopyrite, which he had Smith test, showed a maximum of eight percent copper, which delighted the young man. He refused to reveal his source, but he lost no time spreading the good news. Before Smith left camp, word came back to him that he had pronounced Finch's rock almost pure copper!

Farther along the road, Smith found Fredonia ". . . like so many of our towns is going down, houses lack paint and have become dilapidated, the military academy has been pulled down, and Col. Hill, its former superintendent is dead." On the other hand: "Roanoke appears to be rather more thriving than most of the little country towns of our unfortunate state and I noticed two new dwelling houses of tasteful appearance and a new college building in course of erection."

On October 6, Mr. Sims took Smith to a spot where he filled his pockets with quartz crystals "of varying perfection." Upon reaching Opelika, the president of the Selma & Mobile Railroad invited Smith to ride with him on the cowcatcher to study several interesting cuts in his road. Tempting though it was, Smith chose instead to tour the Chewacla Lime Works at the site of an old quarry. There he had unqualified praise for the enlightened management of the long-time superintendent, Mr. Clapp, who had three lime burners functioning. Two were lined with New Jersey firebrick, the third with Mr. Billingslea's soapstone from Oak Bowery. The latter was unquestionably superior, Clapp said, being the best fireproof material he knew.

The lime was shipped in barrels made at the works, each barrel containing two and a half bushels and selling for a dollar. Combining the agricultural lime, which sold for only fifteen cents a bushel, with other progressive measures, Clapp had in five years brought the worn-out fields surrounding the lime works from an annual production of eight bushels of corn per acre to thirty-eight and a half bushels.

Moving on to Auburn by October 11, Smith stopped off to meet the faculty of the newly founded (1872) Agricultural and Mechanical College and to be entertained by Professor William C. Stubbs; then on toward Walnut Hill and west of it to Ufola, a deserted Indian village. Along the way he stopped over with Parson Guthrie and his twenty children "most of whom were at home. I have seen bedlam on a small scale but never on a large one before," Smith wrote in his journal.

The morning of October 14, Parson Guthrie went with Smith to see the old Silver Hill gold mine region, Tallapoosa County, where, before the war, a company had installed expensive works. Now only the wreckage of the two mills remained surrounded by collapsed shafts and tunnels and an enormous pile of gold-bearing quartz.

On the way to Tallassee, a sudden fit seized one of the horses as it descended a steep pine straw covered hill. The animal leaped about stiff-legged, panicking the other horse and causing the wagon to careen around a sharp curve on two wheels. Fortunately, the wagon righted itself, the driver regained control, and they safely reached their destination which Smith admired as "one of the most attractive and picturesque spots in the State."

Smith submitted his "Report of Progress" for the Geological Survey 1874 to Governor George Smith Houston, who had served twenty years in the United States Congress before the Civil War. When he assumed the governorship, Houston was faced with a poverty-stricken, bankrupt, and devastated state. By the end of his second term, he had reduced taxes and brought state expenditures under control. Smith assured the governor that he understood the financial hardships of the state. However, one of the most important means for restoring Alabama to prosperity would be to provide funds for a thorough study of the state's natural resources by a large-enough work force to accomplish the work quickly.

In his report, Smith wrote of the beds of excellent building stones such as granite, gneiss, and marble in the various counties and the successful uses he had seen made of them, but pointed out that inaccessibility hampered the quarrying and marketing of the stone. The extensive belts of soapstone could be used for "fireplaces, . . . culinary utensils, rollers in cotton factories, . . . whilst for floors of baker's ovens it deserves special mention." He recalled the "fine white porcelain clay" near Louina and Socopatoy and Notasulga; the large deposits of asbestos in Tallapoosa County that had been ignored until lately when experiments had begun to create a textile fabric by weaving long strands of asbestos with cotton and wool.

Though several counties had good sources of mica, Smith wrote, Chilton had the finest specimens. He described the splitting of the large boulders into plates, then the trimming of the plates for the manufacture of lampshades, stove doors, and window frames. The plates sold for fifty cents to three dollars each, depending on size and quality.

The corundum of Tallapoosa County, the graphite of Clay, Tallapoosa, Randolph, and Chambers counties, the gold of Arbacoochee, Old Pinetucky, and Chulifinnee, the copper of Cleburne County all had commercial possibilities. The pyrite contained some copper, admittedly not enough for profit, but "utilizing the sulphur, which so often is allowed to go to waste, and the valuable residue of iron remaining, can be worthwhile."

Smith's summary of what he saw on his tour of 1874 was like a pep talk to the people of Alabama, a forecast of a rosy economic future, filled with a plenitude of opportunities, a plethora of riches—if only the people could catch his vision, if only they would discard unrealistic dreams, if only they would take the long range view and work toward the goal, if only they were open to learning, if only the railroads could be brought in and the rivers made navigable, if only . . .

1875

Bibb, Calhoun, Cleburne, Fayette, Shelby, and Talladega Counties

In the 1875 Geological Survey team, Smith included students from Professor Stubbs's class at the Auburn college. Whereas the 1874 tour stressed to some extent the metamorphic formations, he planned this year to emphasize the Silurian rocks, the sedimentary deposits. On August 15 the team traveled along the high ridge separating Big Sandy Creek from Big Hurricane Creek to Centreville where they set up camp. Here they studied the rocks that created the last falls on the Cahaba River and visited the quarry that produced the white limestone for the 1849 Porter King house, considered the finest mansion in Marion. Besides rocks, Smith collected two snakes, a black and a garter, for additions to the school's assortment of reptiles, first begun by Michael Tuomey.

On August 17 just west of Brierfield, Colonel Smith, a descendent of original Bibb County settlers, guided the men to a coal pit that he had operated during the war. Though water in the pit prevented a study of its layers, Professor Smith observed that the coal piled high around the opening still burned well after eleven years of weathering.

Two days later, while several of the party went deer hunting, the others crossed the Cahaba River to Mr. Owens's to collect saddlebags full of choice specimens until noon. Moving along toward Lily Shoals, they located a comfortable house to stay in, fortunately for Smith as he fell sick and had to remain in camp for a day. They drove on to Montevallo, camping at first in an outbuilding. Later they got permission to stay in rooms above the Central Iron Works' store.

The last two days of August the group collected from the abundance of coal plants in the shales of the Helena mine and at the Alabama Mining and Manufacturing Company above Helena. Although the mines were situated relatively close together, they yielded entirely different fossil plants. Smith made haste to pack two boxes of the specimens to send to Professor Leo Lesquenux, a noted authority on fossil plants found in North American coal measures, for identification. On the way to Siluria, they jounced over a "peculiar ridgy" formation that looked as if "fingers had been drawn over it when soft and plastic."

On September 9, they made the acquaintance of Squire Bushbee who claimed he smelted silver and lead out of "greenish slates" dug from his front yard, an astonishing accomplishment considering that the slates contained nothing, Smith found, except small amounts of iron.

On September 13, writing from the Shelby Iron Works, Smith described the unusual "Iron Mountain," or "The Mountain" as locals called it, east of

Columbiana. It had "the shape of a horseshoe with the two free ends pointing NE. Around the NW arm of the horseshoe flows Beeswax Creek which cuts the SE arm in two at Leeper's Mill." Continuing in the area of the iron works, on the fourteenth Smith discovered fossils of "brachiopod shells, one pentremite, several . . . corals, and two or three trilobites, also great numbers of lace like broyozoans" all weathered from brown shale.

On the way to Fayetteville on September 15, the travelers passed a source of whetstones for Indians in times past. Two days later they heard about an ancient Indian burial site on Bates' Island in the Coosa River near Paint Creek "where the bodies have been placed in pots with another pot over each one."

Smith summed up Childersburg as "a little town located in an old field without any shade trees" whose citizens were looking forward to the coming of the railroad. Onward the wagon rumbled toward Talladega past ditches and fields covered with fragments of promising limonite (brown ore), "rounded and smooth as though water worn." Smith believed a furnace established here should run for years. Rain began falling, and the group took refuge in a wagon yard the night of September 25.

Northwest of Talladega rose Bald Mountain, denuded of trees some years before by a hurricane. The men tackled old Bald on horseback, but it was so steep their saddle girths broke; Smith alone continued the climb on foot. What he saw from the highest point impressed him: "I had an unobstructed view of the whole country from the East around by South to West. Last lay the valley with Talladega in the midst and bounded by metamorphic hills. To the south I could see down towards Syllacauga [the correct spelling at that time] over the Pope Mountains and other ridges which we crossed last week." Using his considerable artistic talent, Smith entered several sketches of this area in his journal, including Talladega from a distance, and later Mt. Parnassus as seen from Cold Water Mountain sixteen miles away. They found a very fine chalybeate spring at Chocco, where Mr. Thompson, the owner, had recently erected several buildings with paying visitors in mind. Smith could foresee a prosperous future for such a health resort.

On October 1 and 2, Smith and John Oden took a long hike through the Kahatchee Hills, crossing Kahatchee Creek near a camping place, Smith remembered, of the University of Alabama cadets when he was in command of them in 1864. At one of the several camps they established back then the cadets stayed for a month, afflicted by "a serious outbreak of pernicious fever," as one historian wrote, but Smith did not dwell on the past. He took pleasure in the abundant water of the Kahatchee Hills and the creeks with their historic names, the Tallasseehatchee, the Kataula, the Wewoka, the Kahatchee, the Emaughee. Later the geological group "rode the entire length of Kahatchee [Creek] to its source in the mountains . . . Night overtook us before we reached the highest summit which I regretted but expect to go up tomorrow," Smith wrote in his journal.

Smith described Talladega Creek in more detail this year than he had in 1874. He wrote of the deep gorges (one hundred to two hundred feet deep) where great chunks of conglomerate had slid down into the water "and lie there now like huge green houses . . . beautiful in color and structure, the grains of quartz showing finely against the pure green of the rock." A mile above Dr. Taylor's mill "the waters of the creek fall . . . in several beautiful cascades ten feet and more in height. The long continued action of the waters has worn the rock into numerous potholes in which flourish many varieties of water plants and numerous fresh water gastropods . . . A visit to this place on the creek and many others above would well repay the trouble and fatigue."

For Smith, all of this beauty translated into almost unlimited waterpower. However, he wrote in his report to the governor: "The rugged nature of the hills which lie between this magnificent water power and the outside world will stand in the way of its utilization for some time to come." He noted that the old Eagle Forge and the Rob Roy Forge had been situated on this creek. Old timers recalled that the Eagle Forge was built from timbers hauled in by a wagon with wheels made of discs sawed out of a huge black gum tree. They remembered the measured pounding of the heavy hammer, powered by the water wheel, as it beat impurities out of the red-hot iron. On still days and nights the sound could be heard five miles away.

At this time, Senator Cunningham and others operated a gold mine on the Talladega. Smith judged that it paid very well based on the sample assays taken at different levels. In addition, he could see gold in the quartz piled around the mine opening. It was evident though that the "piney woods way" that the mine was being worked resulted in the loss of much gold.

By October 7, the geological group set up camp on the creek at the Alabama Furnace, the only active blast furnace in Talladega County. It stood on

Remains of the old Eagle Catalan Forge (1846) on scenic Talladega Creek. Photo by Smith's assistant, Robert Hodges.

the ruins of an earlier furnace destroyed during the war. Though small, with only a forty-one-foot stack, it produced well—"twenty to twenty-two tons of foundry iron per day." Factors contributing to its success were the nearness of the brown hematite ore beds and the source of the limestone used for flux, both about one-half mile distant. Smith believed that the superintendent overseeing almost every step in the production of the iron, even to the burning of the charcoal, was another advantage.

That afternoon they rode to the Cold Water Mountains, a "chain of three or four peaks arising . . . at the Boiling Spring and running a little N of E and terminating at Oxford." On one of the mountains, two extraordinary sinks had formed. The one on the southeast side, nearly three hundred feet deep and several hundred yards across, had in it a running stream falling over the cliffs and somehow disappearing. "A narrow sharp ridge" separated this funnel from a matching one on the opposite side of the mountain. It had no running stream, but must have had drainage as no rainwater had collected in the bottom. Smith decided that the water-soluble dolomite had leached out of the brownish chert composing the mountainside, creating the two strange formations.

On their return trip, the explorers "descended the mountain at one of the steepest and rockiest places, having to abandon the trail as leading us too far towards the NW. Found great difficulty in getting down as we had to lead our horses down a place too steep for them to go except by making zig-zags."

At the eastern end of the Cold Water chain, near Oxford, "between the . . . fingers of the Choccolocco Mountain are the repositories of some of the best of the limonites which supply the Woodstock Furnace." At Snow's Branch lay "a bed of brown iron ore in irregularly shaped lumps which when broken open, show an unchanged kernel of pyrite within, and every step of the progressive change from pyrite to limonite may be observed on a hand specimen."

At the west end of the Cold Water Mountains, a great many strong springs joined together to create the three-mile long Cold Water Creek that flowed into the Choccolocco. "This is one of the most remarkable streams I have yet seen," Smith wrote of Cold Water Creek, "both as to volume of water and its clearness, and also its very short course." He made note of a schoolhouse and a church "beautifully situated" on its banks.

He also praised Choccolocco Creek, "a bold full stream" that had a twenty-five-foot waterfall at Jackson Shoals and sandstone ledges that would lend themselves to easy dam building. "To enumerate all the beautiful springs with which the county abounds would be impossible," Smith recorded.

Between some of the mountains appeared fertile red soil where farms flourished, their fruit and growing crops protected from damaging weather changes and overlooked by "fine plantations and elegant villas." In contrast, the mountains, which predominated, were piled with huge fragments of sandstone. In this bleak environment only a few creatures eked out a living among the stunted oaks—the rattlesnake, the "black adder," and a hard-shelled black and red grasshopper.

"The whole country about Oxford is one of great interest and to map it out exactly would take several weeks instead of days," Smith wrote. "The usual stories of silver mines . . . are current in the neighborhood of all the high peaks of the range, and a man soon makes himself obnoxious by throwing cold water upon these high expectations and visions of sudden wealth." The very next day, October 12, he spent his time in Oxford making "himself obnoxious" with his mineral analyses. "The majority of the people bringing ores . . . have no faith whatever in my representations, being all wedded to the idea that they have silver, copper, etc. on their lands."

The morning of the thirteenth, the party made several mine visits to collect additional specimens. Afterwards, in a drizzling rain, they started out of the mountains with Smith walking the entire distance to Arbacoochee, searching the ground with his keen eyes for a small piece of iron they had lost off the wagon the day they came up the mountain. Upon arrival in Arbacoochee, they again accepted the "jolly old bachelor's" offer of hospitality with less happy results than before. "Old Uncle Billy was very tight and was somewhat annoying on account of his persistent talking the same thing over and over," according to Smith.

From White Plains, they headed over the mountain for Jacksonville. They had not gone far when the wagon pole broke. The two hours required to make repairs caused them to blunder into Jacksonville after dark "but we went at once to Judge Walker's and were there taken good care of."

1876

Bibb, Franklin, Shelby, and Tuscaloosa Counties

July 21 of this year Dr. Smith, taking with him Professor Stubbs from Auburn and Professor R. B. Fulton from the University of Mississippi, went into the field by way of Big Sandy Creek (just beyond Coaling in Tuscaloosa County) over a road "safe only for goats." With Dr. Clements the men looked over the coal openings near Clements Station, the coal beds along Lye Branch, and a lime kiln on Big Sandy Creek. The "several beautiful and very bold springs [that] burst up from below the dolomite" impressed Smith, especially the Blue Spring near the Clements house. The latter he described as having a basin "ten feet square and quite deep. The water boils up boldly, is perfectly clear and has the blue color which gives the name to the spring. Water weeds cover the bottom and sides of the basin." The creek cut through ledges, leaving chunks of rock strewn about, at a place called The Quarry.

They chose to camp near several abundant springs at the site of a historic stagecoach stand. Smith could foresee this area becoming a popular health resort if shelters were built and the springs cleared out.

On July 24, camped at Vance's Station, the group explored an extensive cave "studded" with fine stalactites and stalagmites that had been blackened by the torches of sightseers. Smith broke off one of them to admire its "handsome crystalline structure." That same day they visited Eight Acre Rock, forty acres of hard sandstone characterized by sparse vegetation, wide and deep cracks, depressions filled with white sand, and littered "with pebbles the size of peas." On its western edge they found a large square cave used by the local sheep for shelter.

The geologists proceeded to Bibbville to see a pit furnishing a superior clay for fire brick used in building rolling mills and furnaces. At Greenpond they examined ore banks from which Mr. Edwards excavated a high grade of iron ore that he cleaned in his own washing machine before shipping to the furnace at Oxmoor.

On August 1, the party camped in Jonesboro, having come from Tannehill over a chert road that Smith rated as one of the best in the state. They had dinner at the Pleasant Hill Academy. Afterwards Professor Fulton and Smith hiked out to examine a range of sandstone and some ore beds. Smith surmised that a small ridge just east of Red Mountain would contain red ore, but hard rains and the resulting wet bushes, hampered his checking this. Then on August 7, the deplorable road caused them to arrive late in Helena. Perhaps a nearly perfect geologist's day on August 9 made up for those recent setbacks. "This day's trip has thrown more light upon the structure of the Coosa coal fields than I had thought possible to be accomplished in one day. Returning we went out to Maj. Wagner's quarries and collected specimens for cabinet and analysis. After dark we took tea and spent a pleasant evening with Mr. Thad G. Holt." On August 11, professors Fulton and Stubbs left the expedition, and Smith was pleased to acquire specimens of "very handsome" oolitic limestone from a mine.

At Blake's Ferry, quicksand had washed into the usual landing on the opposite side making the crossing hazardous. The ferryman did not choose the new landing—a fifteen-foot-high bluff with a forty-five degree slope—with mules in mind, especially mules pulling a wagon loaded with camp equipment and rocks. During the struggle to climb up from the ferry, the wagon shot backward crashing into the ferry and breaking its chain, then rolling body-deep into the river. All the equipment and the heavy collection of specimens had to be unloaded before the wagon could be pulled to safety. It was eight o'clock that night before the geologists succeeded in setting up camp atop the dry bluff—they had to paddle a distance upriver in a dugout to fetch water.

Near the close of the 1876 summer trip, a nameless branch flowing into Buck Creek in the vicinity of Siluria moved Smith to poetic praise: "There are few scenes more worthy of the pencil of an artist than those found wherever the Millstone Grit has been broken through by a comparatively insignificant branch. The tough hard rock yields only enough to allow the water to pass through making deep, narrow gorges with towering precipitous cliffs of sandstone on each side."

1877–78

BIBB, BLOUNT, COLBERT, CULLMAN, FAYETTE, FRANKLIN, LIMESTONE, MADISON, MARION, MARSHALL, RANDOLPH, AND WALKER COUNTIES

Mid-July 1877 found Smith, accompanied by his father, Dr. Samuel Parrish Smith, once again at Dr. Clements's place on Big Sandy Creek where they took dinner and rested. Soon after leaving there, with a heavy rain coming on, their hired horse balked. They lost two hours while Anderson, Smith's driver, fetched a mule borrowed from Mr. Burns, enabling them to plod on to Burns's house where they put up for the night. Next day Smith sent the contrary horse back to Tuscaloosa and rented a mule. Proceeding to Scottsville on Schultz's Creek, Smith acknowledged to his journal, not for the first time and not for the last, "The relations of the rocks here are extremely obscure and I must confess that I have not the key to the puzzle."

A week into the trip "Pa" returned home, leaving Smith with regrets that due to heavy rains and several accidents they had been able to work only a day and a half while he was with the expedition. Onward to Pratt's Ferry, thankful to escape the flea infestation at Scottsville, they found a vacant house for camp. Smith collected a forty-pound chunk of barite from a large bed, went in Mr. Wallace's skiff to visit a limestone bank where many varieties of ferns flourished (three of them he had never seen before), and clambered through a dense thicket to reach an ore bed. Smith could not identify the plant that made up this "almost impenetrable" thicket except for the local name of "privy plant" but he commented on its "peculiar aroma."

July 29 was such a scorcher he stayed in camp till late afternoon when he rode to Mr. Hansberger's, postmaster at Termus, to arrange a trip to Six Mile Creek. Two days later, the two men rode to the place where the creek went under ground, "the sink," and continued along its dry bed until the water reappeared.

Heading homeward Smith retraced his route to the Clements place pausing to look at "the blue hole," a crater-like pond situated on a high hill. Locals reported that the water level remained constant in this "bottomless" pond. The 1877 geological expedition wagon returned to Tuscaloosa in early August in time for the birth of Julia Allyn Smith on August 7. The baby was a namesake of Eugene's sister, Julia Pratt, who was married to Merrill Pratt Sr., nephew and close business associate of Daniel Pratt.

A year later, on August 6, the wagon headed for the counties north of Tuscaloosa to begin a firsthand evaluation of the coal resources. Along the way, eighteen miles out of Tuscaloosa, Smith tracked down the rumor of a six- to eight-foot-thick seam of coal, which turned out to be, according to the locals, either nonexistent or perhaps eight to ten inches thick.

Smith judged that the forest growth ("red, white, post, chestnut and Spanish oaks, a few short leaf pines, some hickories, chestnuts, black gums, and tulip trees") on the high ridges they traversed denoted fertile soil, but Mr. Hamner, a native, characterized the shallow soil as "very drowthy" because of its inability to retain water with a resulting high rate of timber mortality. Smith noted that along the summit of the "sinuous ridge" separating the two main drainage systems of the Tombigbee and the Warrior rivers ran the historic Byler toll road, a turnpike opened by John Byler in November 1822 to link Tuscaloosa and north Alabama.

Heavy rains three days running forced them to move their camp into the new Walker County Courthouse "where we were much more comfortable," Smith wrote. By the fifteenth, the expedition was in the Arkadelphia area of Cullman County, where they had to backtrack seven miles to recapture the mules that escaped the previous night. At the confluence of the Sipsey and

Mulberry forks of the Black Warrior River, they passed through Old Warrior Town, "the scene of the death of Black Warrior," Smith wrote in his journal. The site was "beautifully situated," as was typical of Indian settlements, on a high bluff with a far ranging view to the north. Here the geologists crossed the river by ferry, and followed the Mulberry fork northward. At the Gap, they descended two hundred feet to the Mulberry Bottom with Smith marveling at the beauty of the surroundings. One early morning as he viewed the Warrior basin toward the east from the top of a 450-foot bluff, "a line of white fog revealed the track of a little river, which drains this part of the basin."

Here in the coal measures, Smith met with the usual impractical dreamers and the ever-recurring tales of lost Indian silver and lead mines. "They all believe they have . . . mines of silver, lead, and copper to say nothing of iron and coal. The silver and copper have thus far proven to be iron pyrites."

About ten miles from Jasper, they visited Clear Creek falls where the water "pours over two bluffs of sandstone 30 feet high each and about 200-300 yards from each other . . . It is almost impossible to get down to the foot of the first falls on account of the precipitous bluffs, but with trouble we can get down to the water level below the lower falls. If this place was improved I don't know any more beautiful and attractive scene."

On August 27, Smith wrote of a coal bed two and a half feet thick worked by Ira Moore on a branch draining into New River. It yielded "a fine quality of gas coal" that was loaded onto barges during high water, floated downriver, and sold in Mobile for $2 a ton.

After leaving Mr. Moore's, a prolonged downpour caught the geology wagon and the buggy accompanying it; during the hardest part of the rain the buggy's harness broke and had to be, not repaired, but "patched." Night found them still on the road, but they slogged along till they reached the nearest hotel. Rain restricted Smith for a few days, but he was "very glad of a rest, and I have no doubt the mules appreciate it still more for we drove them about 30 miles and part of the time over the worst road we have seen."

He took a side trip to see a source of a "peculiar substance" called cannel coal but the excavation had been refilled and nothing of interest could be seen. However, a man agreed to send Smith samples at his camp in Fayette. In this area, Smith explored rugged Jaggers Branch at its joining with Mill Creek, but progress was difficult because of the numerous huge rocks that had rolled down into the streambed. Smith found that the thickness of the Jaggers coal bed had been greatly exaggerated; it did not exceed four feet.

Continuing in the vicinity of Fayette, he found a second source of cannel coal that overlaid, as it did at the previous location, the common bituminous coal. The man who collected specimens from the first site told Smith he asked "Old Man Porter on whose land the coal occurs . . . if the Geologist could have a little of it. 'Yes,' said he, 'he can have a little, but I'll be damned if he can make a practice of getting it.'"

In his journal, Smith continued extolling the wild, rugged terrain—the steep hills, deep ravines, and hundred-foot high sandstone cliffs along the creeks. Under the overhanging rocks, the caves sometimes extended twenty-five feet deep with ceilings reaching even higher. The damp walls nurtured rare and beautiful ferns, a botanist's delight, and from many of them flowed "springs of delightfully cold water, sometimes strongly impregnated with iron."

On Raven's Nest Branch near Pikeville, Smith found some of the most beautiful clusters of native ferns that he had ever seen: lobed spleenwort, mountain spleenwort, wall-rue spleenwort, Peters' fern, and *Trichomates radicans* [no common name] with fronds six to eight inches long. Peters' fern, the smallest fern in Alabama, "was discovered by Judge T. M. Peters of Moulton in the rock houses . . . of the headwaters of the Sipsey Fork of the Warrior," Smith wrote.

Two years later, Judge Peters made a valuable contribution to the rebuilding of the university library's lost resources. In addition to rare books on botany, he donated his own collection of plants that he had carefully preserved and mounted. His letter accompanying the presentation was written in Latin. Smith and another faculty member were delegated to acknowledge the gift, "in Latin, on parchment stamped with the seal of the University," according to Dr. James B. Sellers.

Also near Pikeville, Smith found evidence of gold in a flow of large quartz rocks that he believed had washed from western North Carolina by way of the Tennessee Valley into Marion and Fayette counties. He looked over the "hard and massive" cap of Blaylock's Mountain that furnished millstones for a demanding market. Continuing northwest, Smith appraised Lem Burnett's coal that he extracted from a creek bed. It was highly prized by blacksmiths who traveled far to get it.

The hazards of the trip were not always caused by the mules or the roads as noted in Smith's journal entry during the first week of September: "Stayed all night with Mr. Shirley, who was not prepared to entertain people so he said. I found to my cost that he was telling the truth, for I woke with a violent headache, and on my way home could scarcely sit up from nausea and sick headache, brought on from the cornbread and bacon eaten last night and this morning."

By September 10, Smith had made it to Frankfort, "a dilapidated town" in Franklin County. At the springs there he found the ruins of several cabins, a spacious dining room, and under an overhanging rock above the chalybeate spring, a tenpin alley. The geologists camped on a bluff so steep they could not get the horses down to the water for drinking, but had to lead them a quarter mile along the road to a puddle.

Near Barton, Smith examined six rock strata on a hill behind the cotton factory. Two of them he found "peculiar." He decided that all of the high cliffs including those on Cane Creek, Buzzard Creek, and Rock Creek merited several days' study. Enduring "cold, raw, windy" and sometimes rainy weather, the geologists welcomed the railroad agent's offer of "comfortable quarters with a good coal fire in the Depot building."

The predominating limestone of north Alabama had eroded into numerous caves that sheltered thousands of bats with a consequent layering of guano on the floors. Smith had Henry McCalley, his able assistant who worked for no salary, write a treatise on the guano and its value for fertilizer to include in his later report to the governor.

On September 13, Smith dined in Tuscumbia and looked over Mr. Pybas's fossil collection including stem pieces and a stump of lepidodendron from the well-known fossil beds of nearby LaGrange. The survey crew planned to camp that night at LaGrange but had to go beyond it to have access to water. Before the war when Tuomey visited the town, it included an academy, a preparatory school for boys, and a college and military school besides substantial churches, a Masonic hall, many lovely homes and the usual businesses—dry goods stores, grocery stores, a woodworking shop, a blacksmith shop, a tannery, a tailor's shop, and a post office. But on April 28, 1863, the 10th Missouri Cavalry totally destroyed the town. As Smith wrote in his journal, LaGrange was now "most truly a deserted village" with only a couple of ruined houses remaining, one

The old cemetery in the once-thriving town of LaGrange, Alabama.

of them occupied by the widow of the college president, J. W. Hardy.

Next day the group browsed about the site harvesting a large number of lepidodendron fossils and sigillaria fragments. "Saw the large trunk described by Prof. Tuomey and two large stumps and another trunk of lepidodendron by the side of the road in front of what used to be the house of the President of the College," Smith wrote.

"All buildings are now down and the place looks desolate . . . Gave [the] . . . widow of Prof. Hardy, $10.00 for the lepidodendron tree." Smith paid Mr. Pybas $10 to haul the fossil tree down to the railroad at Leighton, where Smith had already packed and shipped the smaller specimens they collected.

Professor Hardy had begun his extensive fossil collection about 1853. The specimen most prized by Smith, "about three feet in diameter and four feet high, showed distinctly the leaf impressions . . . [T]he sandstone of which it is composed is full of impressions of other plants, calamites, etc. These . . . appear to show that in the process of petrifaction, the interior of the trunk was removed by decay or otherwise, leaving a hollow cylinder of the outer layers of the trunk (bark perhaps), and this hollow cylinder was filled up with sand and fragments of calamites and other coal plants. The subsequent induration of the sand and the removal of the vegetable matter which formed the mould would give the fossil as we now find it. Another specimen showing the leaf scars imperfectly is remarkable for the reason that it is the lower part of a trunk about the size of the preceding, but with two large roots attached . . . All the gullies which furrow the sides of the hill below the town on either side are constantly exposing fragments of these remarkable plants."

From Leighton, the geologists made their way to Courtland, where Smith spent Sunday morning, September 15, in camp writing letters. In the afternoon, he rode out in the countryside with Professor Stubbs. Next day Professor and Mrs. Stubbs, another couple and Smith, conveyed in a dirt car pulled by an engine and guided by Major J. C. Long of the United States Corps of Engineers, toured the new locks on the Tennessee River. "These are splendid pieces of work, made of cut stone, . . . the crystalline limestone . . . found some four or five miles up the river," Smith wrote. They also went to see one of the old locks still standing.

On September 18, the geologists crossed the Elk River, leaving behind the "red soil, cleared fields, old settled places, and beautiful farms" and entering the "moderately broken cherty, very little cleared up country" on the way to Athens. They traveled through a corner of the Barrens where the "yellowish white soil" supported mainly oaks—black jack, post, and red—reaching Huntsville by nightfall. On the nineteenth, Smith took care of business in Huntsville and visited Monte Sano where everyone "drank from the Cold Spring which issues from beneath one of the topmost strata of limestone."

In Morgan County six miles south of Whitesburg and five hundred feet above the Tennessee River, they camped amid Indian kitchen middens then crossed the river on a flatboat, which Smith, an expert on ferries, termed "first rate." By Sunday, September 22, they were in Blountsville, known as "Bearmeat Cabin" in the days when it was headquarters for an Indian chief. They had traveled there from Guntersville by way of Big Spring, where Smith admired a "sheet of water 30 feet or more in diameter," which, he discovered, was located incorrectly on the map, as was the crossroads town of Brooksville ten miles from Blountsville.

Having the mules shod, and a heavy downpour of rain with consequent rising waters, delayed them September 23. Dry Creek was so out of bounds that they camped in a vacant hut waiting for the water level to drop. They chose to cross the Mulberry Fork and the Locust Fork separately rather than tackle them after they joined to form the Black Warrior River because of the latter's dangerously high waters.

Smith submitted the combined 1877–78 Geological Survey of Alabama Report to Governor Rufus Cobb. Again he expressed concern about the lack of funding for the survey and again he stressed the importance of the state's natural resources to its economic health. He pointed out that evaluating the coalfields was more difficult and expensive than evaluating other geological formations. The sum allowed him by the legislature, he wrote, was barely enough "to pay for ordinary expenses leaving nothing for such work as would be absolutely necessary for the thorough exploration of a coal-field." In his preface to the report, Smith noted that the "maps accompanying this report [were] taken mainly from the large State map of Col. D. H. Cram" with the colonel's permission.

1879

WARRIOR RIVER

Smith's summer trip of 1879, under the auspices of the War Department, was for the purpose of evaluating the Warrior River: to survey it, to explore the feasibility of making it a year-round navigable stream, to estimate the cost of such an undertaking, and to present an inventory of natural resources that would be made accessible by opening up the river. At that time, river commerce was at a standstill due to "the uncertainty as to the time and sufficiency of high water and the shooting of the falls, in particular those near the foot of Squaw Shoals, being attended with much danger to property and some risk of life."

Smith believed if the river, which he estimated covered about five thousand square miles, was made navigable, the coal trade would flourish. On this tour, however, he had to gather information that would show whether the results would be worth the cost of the improvements. The expedition party, including Henry McCalley, entered the coalfield from a different direction this year, and got underway in adverse circumstances—an all-day downpour on their camp in a Jonesboro churchyard, and a disabling headache for Smith. He briefly noted in his journal that they were in the area of Five Mile Creek that "flows parallel with Rock Mountain to its confluence with Valley Creek." By August 4, still at Jonesboro, he entered many sketches in his journal showing rock layers and uprisings, and collected ferns and rocks to send to Tuscaloosa.

From the Pratt mines, flourishing under the leadership of three giants of Birmingham industry—Truman Aldrich, Henry DeBardeleben, and James Sloss, Smith wrote on August 16 that they were waiting to hear how they should proceed with surveying the river. Three days later, joined by Joseph Squire and three "hands," they went on foot to Calvary Williams's to measure two coal seams. From there they proceeded by way of the Short Mill road that was "next to impassible. If we had not had extra hands we could never have made it."

Camping at Snow's Ferry on August 29, Smith drew a diagram of nearby Toadvine and described the town: "It consists of two or three stores, about as many houses, and a photographer's tent." The postmaster, C. L. Smith, also ran a gun shop, and his brother had a smithy and a wagon shop. From C. L. Smith, Professor Smith collected several "Indian relics, amongst them a peculiar pipe of steatite." On this same day, Henry McCalley fell in the river getting nothing worse than a "thorough wetting." Smith kept busy in camp writing out instructions for the surveyors who had set up their own camp to suit their purposes.

On September 3 in the vicinity of Lost Creek and Fanny's Branch, the exploring party was surprised by "an unexpected occurrence of Drift beds, quartz pebbles, ferruginous sandstone and conglomerate," instead of the usual "fragments of chert and flint." A week later not far from the Dent post office, they "enjoyed a luxury for this part of the world, viz. a draught of water from a freely flowing spring." After dinner that same day, they rode to Shepard's Bend and crossed the river looking for the surveyors only to find they had moved to Sander's Ferry. Smith and his party continued upriver to Davis Ferry where they stayed the night in a small hut. Along the way they observed several coal openings in Barton's Creek near the river.

Smith took the next day off, staying in camp to answer letters and to draft additional instructions to the surveyors. On August 11, he examined "the old shaft on the bank of Lost Creek below the ford. Coal lying at the mouth of the pit is good, hard, and bright, excellent quality." In the fork of Lost Creek and Cane Creek, on land owned by John Gomer near Burnt Meeting House, was another bed of similar coal that had been mined extensively. It had "sufficient bony coal in it to make it a splendid stocking coal and not enough to injure it as a fuel. The coal is well praised by black smiths."

On September 13, after a night of rain, the mules were unable to climb a slick hill on the west side of Lost Creek. The men had to "take out picks and roughen the ground" to provide the mules a foothold. That night they camped on the road to Jasper in an old church they knew from previous tours. Smith spent most of the night of September 17 taking care of the large amount of mail that awaited him in Jasper. He usually wrote detailed letters, sometimes running to six or seven handwritten pages and dealing with a wide variety of subjects.

The geological team next toured the Flat Creek and Horse Creek coal sources that produced hard, bright coal almost free from bone. Smith likened it to anthracite and wrote that it did not "smut like ordinary bituminous coals." After the war, many boatloads of this coal had been shipped by way of the Warrior River to Mobile for gas making, but the geologists saw a place where several loaded boats, after waiting years for a tide to set them afloat, eventually capsized and sank.

The party came out the Oaky Hollow Road and camped across the river on September 19 where fever and severe headache plagued Smith. The following day, they set up camp in an empty house at Toadvine. Word reached them that the surveyors had been delayed at Snow's Landing waiting for supplies from Birmingham.

By September 23, Smith was confined to Wash Maddox's house fighting the chills, but Mr. Squire, an Englishman experienced in the coalfields and well qualified, inspected several coal works in the area. Three days later, Smith and Squire left Toadvine to go downriver where they dined with Tom Simpson on Indian Creek. From there they rode to Squaw Shoals on the Black Warrior to get "a fine view of them from the top of the bluff which overlooks them. The bluffs are 150–200 feet high and from them we could see nearly the whole length of the shoals since the river at this place has nearly a straight course. There are no bottom lands along the river by the shoals but the water has cut a gorge through the hard rocks and has never made bends to right or left by which bottom lands could be made."

Twenty-six miles above Tuscaloosa, the treacherous Squaw Shoals stretched seven miles upriver through what Smith called "the wildest country I know." Ethel Armes wrote, "About one boat in every eight that passed into the Shoals was lost." One year, out of twelve loaded flatboats that tried to negotiate the shoals, five wrecked. Some of the most experienced river men were numbered among its drowning victims.

From Simpson's they went to Robert Robertson's and "rode again down to see the shoals near the foot of the same. Here there is considerable fall and this is the place where most of the boats have been wrecked." Though Smith's exhaustive survey confirmed the most optimistic possibilities for the development of a lock and dam system on the Warrior River, it did not bear fruit until 1896 when three locks were opened making the river navigable year-round nine miles above Tuscaloosa. Not until the seventeenth lock and dam opened in 1915 was the system completed. However, it was Smith's "groundbreaking" evaluation in 1879 that eventually opened the vast area of coal and iron to the markets of the world.

1880

AUTAUGA, CALHOUN, COFFEE, DALE, DALLAS, ELMORE, HENRY, LEE, LOWNDES, MARENGO, AND WILCOX COUNTIES

Smith was not well during the 1879 expedition. He wrote his long-time friend James E. Graybill in January 1880: "I have this fall been a little overworked and about Christmas broke down and have not been worth killing since." Nevertheless, during January he took charge of installing gas at the university in conjunction with Dr. Peter Bryce at the State Insane Asylum—planning, measuring, sketching, ascertaining prices, ordering parts, and striving to stay within a scanty budget. Sixty rooms, the kitchen, mess hall, and Smith's laboratory, were to be lighted and heated with gas. Freight rates worried Smith—he urged his suppliers to hold the items until all he had ordered could be sent in one package. "The charges for freight on so many packages have already run up the costs to the limit of my appropriation."

Also that month, he began planning his next geological trip. On the twenty-ninth, he wrote Governor Rufus Cobb requesting the $500 annual expense appropriation for the survey. "I should be glad to get the money soon as I have some needed preparation for the following summer's work." In February, he ordered a tent with a center pole "like [a] jointed fishing rod" for easy packing in the wagon.

In 1873 Smith had correctly foreseen the state geologist position as a means to acquire equipment for his university laboratory. Now, for the same reason, he undertook the Alabama-Florida Cotton Culture Report for the 1880 census working with Dr. Hilgard. He immediately purchased a large steam boiler and "a fine chemical balance" ($115) that he expected to remain in his laboratory where "all the analytical work of the whole Cotton Culture Report will be done."

The second week of February, Smith was in south Alabama collecting soil samples for the census report. Back in Tuscaloosa by the nineteenth, he again wrote to Montgomery asking about the $500 summer requisition. He also wrote the tent maker repeating his request that the wooden corner supports and the ten-foot center pole of his tent be made lightweight and telescoping.

During winter and spring of 1880, a threat to abolish state geological surveys and replace them with the United States Geological Survey took up much of Smith's thoughts. He wrote numerous letters and telegrams stating his opposition. To Clarence King, head of the United States Geological Survey, March 5: "I believe that each state ought to keep up its own Geological Survey, and should regret very much to see these surveys discontinued." He explained how Alabama furnished him each year the $500 for expenses and every two years $200 for maps and sections, and nothing to Smith himself. However, he wrote of his hope of seeing "in prosperous times, a properly equipped survey organized, and of having myself some part in it, by which I might by hard work make some credit amongst geologists. I should be sorry to see the opportunity, which seems now very near, removed."

Since he first became state geologist, Smith had made detailed plans of the dream wagon he wanted for his trips. Lately he had been corresponding with Abbot Downing Company about having the wagon built but on March 8 he wrote them: "I find much to my regret that I shall not be able to get a new wagon this spring, and I am very sorry to have occupied so much of your time in getting my questions answered. Be assured that I should not have written if I had not at the time felt certain of ordering the wagon this spring."

In March he ordered for the census work two aneroid barometers from John Browning of London. He specified that the barometers be made to register altitudes no higher than four thousand feet, Alabama's altitudinal limit. Later in the month, Smith did a meticulous proofreading of a manuscript of Alabama plants compiled by Dr. Charles Mohr, a Mobile pharmacist who was also a recognized botanist. The two men knew each other only through their correspondence about plants, but now Mohr was also working for the census. Smith asked him in his travels to collect for the university herbarium. He confided to Mohr his wish to make "a collection of Alabama woods, each specimen to be in the shape of a volume, showing back, cover section, polished and unpolished surface."

Despite all else he had to do, Smith's commitment to the university orchestra continued. On June 7, he inquired of Andre & Co. of Philadelphia, the prices "of ebony tail pieces . . . of violins," and requested a catalog of orchestra music.

In a June letter to Angelo Heilprin at the American Academy of Science, Smith explained he had not time now to send him Professor Tuomey's collections for identification, but hoped he could do so in spring of 1881. "Prof. Tuomey's collections after the destruction of the University during the war, were moved several times by inexperienced persons, who had no idea of the value of a label, and the consequence is that not one-tenth of the specimens can be labeled as to locality."

Also in June, with his summer excursions in mind, Smith wrote the superintendents of the L&N Alabama Railroad and the Alabama Great Southern Railroad requesting free passes. After receiving the passes, he wrote both men promptly, thanking them and enclosing the stamps required for their letters.

Replying to Edward Gillett's request for *Asplenium ebenoides* (ebony spleenwort), a rare hybrid fern with the ability to reproduce itself, Smith wrote that he had "a small quantity . . . growing in my yard upon rocks, and a few

What was known to Smith as Black Creek Falls at Gadsden is now called Noccolula Falls. Photo by Robert Hodges.

specimens prepared, in the herbarium, but not enough to furnish what you desire. Miss Julia Tutwiler of Green Springs, Hale Co., discovered the fern in that locality and she might be induced to collect specimens for you though I think it rather doubtful."

On July 25, Smith wrote Dr. Mohr how he envied him his recent botanizing trip through Autauga County, but Smith himself had just returned from Atlanta with the intention of beginning next day an extensive wagon tour on behalf of his work as special agent of the Tenth Census: to Birmingham, thence to Gadsden, then to Florida. He sent Mohr all he had of the privy plant, urging him to go to Pratt's Ferry to see the plant himself and to visit with Postmaster Hansberger who would assist him in any way he needed.

The end of July found Smith camped in the northern edge of Gadsden renewing numerous friendships, catching up with his correspondence, and making out his reports and financial accounts for the month. In the afternoon of August 1st, he visited Black Creek Falls: "The creek pours over a ledge of sandstone of coal measures about 100 feet high. The overhanging ledge is shaped like a horseshoe with conversity upstream. Under this cliff is a rock house larger than any seen in Winston or those upper counties." In this rock house behind the falls, Smith had the pleasure of finding the bulbless bladder fern (*Zystopteria bulbifera*). A pleasure of a different kind was the bath he took in Colonel Kyle's bathhouse below a spring near the falls.

In the Attalla area of Etowah County, Smith saw much of geological interest but could spare no time for investigation because his focus this trip was agriculture. As they approached Jacksonville on August 3, the day after election, they "met dozens of drunken men, and even now [at] ten p.m. the shouts from the bar rooms make night unpleasant to the rest of us." The wagon arrived at the Alabama Furnace in the rain. They were offered a cabin, "a great favor as we should have had a wet and disagreeable time of it camping out. Colonel Gladden with his usual courtesy invited me to stay at his house and put the mules in his stable."

In Wetumpka, Smith could not resist collecting specimens of tantalite "from the only American locality as yet certainly known . . . I had the pleasure of finding a piece imbedded in a fragment of quartz." In addition he collected crystals of tin stone. "The locality is well worth a more detailed examination which I hope to give it some of these days," he wrote. From the Wetumpka area, they took the Jackson Trace which ran parallel to the Coosa River.

In Montgomery on August 12, Smith left the wagon in the care of Anderson and the blacksmith to have a broken spring mended. He took the train to Opelika, where he hired a conveyance to take him to Mr. Jefferson's. Here he learned about the area's two kinds of soil, the red and the gray. The early settlers prized the red lands, and there was where they built their homes and made their farms. Opinion, however, had changed, and the gray soil was preferred "as being more certain, more easily tilled and even more fertile," Smith was told.

On the fourteenth, he took a sightseeing trip around Tuskegee reporting the town "attractive," and was joined by Anderson and the wagon. Along the road to Abbeville, he noted a great deal of cotton growing. "The soil, however, has in many places washed down from the hills leaving the bare underclay upon which the crops do not thrive." In Abbeville on August 19, he found most of the chimneys "made of aluminous sandstone with casts of shells" taken from a quarry four miles from town. Near Echo, Smith saw two oaks he did not know, "one with large leaves much cut, with wide open sinuses, the other . . . [with] small narrow leaves like the willow oak but the whole appearance of the trees differs from that of any I know."

On August 20, Ozark welcomed the wagon cordially, and Smith visited a friend from the war days, Captain Hughes of the 33rd Alabama Regiment, Infantry. Here again most of the chimneys were made of aluminous sandstone taken from a nearby quarry. Now the wagon headed for Florida, arriving by August 22 at Marianna, a town Smith found depressing due to "the long sandy street with occasional live oaks on each side" and "the general dilapidation." Nevertheless he gloried in the limestone road they took out of town and at the six-mile post was treated to the sight of the largest spring he had ever seen.

Leaving the wagon with Anderson again, Smith struck out southward by rail. He detoured ten miles outside Tallahassee to visit for three hours with Mr. Hubbard, an entomologist. Returning to Tallahassee on August 18, he "came down to Live Oak, so called probably because nothing but pines and not a solitary live oak tree grows any where near the place." On September 1 between St. Augustine and Tocoi, he admired "a fine barren and swamp . . . full of beautiful and to me unknown flowers and tomorrow I shall botanize among them."

In Gainesville on the ninth, Smith met an agent of the land office, a collector of natural history specimens for the Smithsonian. "We drove together out to the Sinks, a remarkable string of deep limesinks probably just above the former subterranean outlet of Payne's Prairie. Until 1871 this was a one to two mile wide (1-3 mi.) open glade . . . through which ran a small stream that emptied into a cave on the N. side. This large low area was destitute of trees, but covered with a fine turf of grass and was a great pasture to which people drove their cattle for miles. The storm of 1871 stopped up the outlet partially so that in one day the prairie was converted into a lake and such it has been ever since."

Smith arrived in Bainbridge, Georgia, on September 13, expecting Anderson to meet him for his return to camp. The train was late, Smith had no breakfast, his head was splitting, and Anderson was not waiting—he had not received Smith's letter. He girded himself and set out walking the five miles to camp where he "was sick enough till about 4 o'clock."

By September 15, they had begun their homeward trek. West of Campbellton, Florida, they crossed a slough "over a bridge that bent under us and that let the mules' feet slip between the springy boards at every step. I was sure we were done for . . . The ferry landing on the other side was bad and we had much to do to make it safe to go down [onto] the flat. The regular ferry man being off on a hunt the duties were performed by two raw hands and I was not at all sure we would get over safely. They had the face to charge $1.50 double the regular fare. I hope to get back 75 cents as Mr. H. Watson at [the] store near the landing said he would report them."

The wagon was once again in long leaf pine and wiregrass country on September 19, traveling through "miles of dreary lonely piney woods" to Captain Kimmey's, six miles from Elba. "Found Capt. K. quite unwell but he remembered me from1862 when we were in the same regiment." On to Elba next day where Smith "turned down the river four mi. and stopped at dinner with L. P. Lee my old companion in the Army."

On September 24, the wagon rolled "into Perdue Hill about five and a half p.m. and stopped all night with my old roommate and friend Jno. L. Marshall." The next day Smith and Marshall came "down to Claiburne and spent an hr or two in getting some idea of the celebrated Bluff . . . The Claiburne Bluff is something over one hundred ft above the river level now." Smith complained on September 25: "The further we come [toward Lower Peachtree in Wilcox County] the worse are the hills, till finally the country seems to be nothing but a mass of small rounded hills like potato hills, and over this the road passes up one and down for several miles. The clay is very deep and much cut up so that we have difficulty in getting along. These are the so-called lime hills."

Before breaking camp, Smith "stepped out to see the piece of ground where the zeuglodon vertebra was found. This was a [small] prairie hill along the sides of which were strewn fragments of yellow limestone with numerous fossils" including sharks' teeth, fish spines and vertebrae. The zeuglodon, a prehistoric whale that grew up to seventy feet in length and lived in the shallow sea that covered southern Alabama and Mississippi forty million years ago, had long interested Smith, but he had had little success in finding the fossils.

On the twenty-sixth, he visited Mr. Burriss, another old friend from Company B. His place was on a ridge dividing the Alabama River on the east from Bassetts Creek on the west. Gullies on both sides of the road were twenty feet or more straight down. Sometimes the gullies came so close the road was narrowed to twenty feet. In the bottom land, Smith found *Pinus glabra* (spruce pine) growing, "some . . . specimens . . . three feet in dia. at base and of correspondingly magnificent proportion." The maximum diameter recorded for this tree was only two and a half feet. Nowhere was it listed as abundant in its range, yet Smith heard it was common in these swamps. [Later in the fall near Linden in Marengo County, he found a plentiful growth of *Pinus glabra*. The natives called it "white pine" and made buckets and tubs out of it.]

Smith stayed over with his old friend Morrisette then proceeded through the river plateau where with the use of fertilizer raising a bale or more of cotton per acre was common, and on to Camden. He met a number of friends there but could not linger. "Left Camden about three, hot as could be, cloudy . . . and after an hour or two it poured down rain till night."

Beyond Pleasant Hill, they crossed "a bottom of the worst kind of mud two and a half to three mi wide." From there they arrived at Benton and set up camp not so very far from Prattville and Smith's extensive network of relatives and friends.

On October 25, Smith sent Dr. Mohr the Latin names of numerous plants he had lately observed. "Cast. Bullifera you have doubtless found yourself, as it grew under the cliff at Black Creek falls and you were there ahead of me. I have it now growing in my yard." He was convinced that the privy plant or croton he discovered at Pratt's Ferry was indeed a new species, he wrote Dr. Mohr. He added that he would soon send him the plants he collected on the Alabama-Florida trip. He added in regard to his work with the census, "Our chemical work is finished and now comes the weary writing of reports."

That same day he wrote Postmaster Hansberger expressing concern over his mishap with a runaway horse. Despite this, however, he still urged Hansberger to collect flowers from the croton plant once a month "getting as many distinct stages . . . as you can find on the bushes." Smith also needed a rooted plant suitable for a flowerpot. He cautioned Hansberger to take great care with the root system and gave him detailed instructions on pressing and drying the cuttings.

In November Smith mailed to Harvard botanist Asa Gray a garden plant with the common name of "shower of diamonds." He enclosed his and Mohr's list of Alabama plants that they had made from their collections, adding, "I expect to use it in making a complete list someday of our native plants."

Smith's unpaid assistant, Henry McCalley, wrote the report on "Alabama North of the Tennessee River" from his travels through Lauderdale, Limestone, Madison, and Jackson counties. He highlighted "the beautiful Tennessee," the many "bold, rapid streams," the "clear, sparkling" waters, "the purest and best drinking water," the lack of lakes and marshes. Of the four counties, in his judgment Jackson was the most interesting geologically. In Limestone he singled out Nubbin Ridge, the only significant crest in the county, for its sterile soil, and wrote that north of the Tennessee he heard the same implausible treasure tales he had heard in lower Alabama. He spent time at Mrs. R. F. Hunter's farm at the mouth of South Little Raccoon Creek where a human skeleton in a clay jar had been unearthed. He admired her exceptional collection of "the large coral Lithostrotion canadense."

1881

Tuscaloosa

Early in his 1881 correspondence, Smith requested from B. Westermann and Company music price lists for an eleven-instrument orchestra with emphasis on "the works of Mozart, Hayden, Beethoven overtures, as well as . . . operatic selections, and opera music." He wrote Mr. Squire, who had accompanied him on the 1879 coal measures trip: "We are driving on here as usual, I spend all my spare time writing up Census and once a week take a turn at music with the orchestra. We play now, not only a better kind of music, but play it better than we could last year at this time." Smith expressed the wish that he, too, could go abroad and tour the coalmines as Squire intended doing.

In February, he ordered music from Martens Brothers, New York, specifying that they not send "more than $15 or $20 worth at once." The orchestra planned to try out the pieces before selecting the ones they wanted to keep. He also requested the lowest price on eight to ten folding music stands. That same month, he wrote Dr. Hilgard: "After much deliberation and several at-

The Smith home on the University of Alabama campus. Smith brought back from his journeys many rare plants to cultivate in his yard and exchange with other botanists.

tacks of the 'mulligrubs'" he had come up with two possible plans for the soil analysis report which he now submitted for approval.

When Smith heard that Professor C. C. Eaton planned a supplement to his fern book, he offered to send Eaton specimens of the *Asplenium ebenoides* he had growing in his yard to provide a better illustration for the new edition. Smith added that he also had some "magnificent specimens" of the Appalachian filmy fern (*Tr. radicans*).

On March 15, Smith wrote James Edward Graybill in New York, "I am just now in a terrible press, trying to get a census report worked off and at the same time to put a State Geological Report through the press." Ever since his return from Florida, he had been working steadily on reports and as yet saw no end to them. Hearing that Graybill was just back from Europe prompted Smith to reread the journal he kept during the time they were students on the Continent after the Civil War. "I live in hopes that I may make the trip again some day," he confided to his friend.

March 27 marked the birth of the Smiths' second son, Landon Garland, named for Jennie's father. Eugene did "duty as housekeeper" while Jennie recovered. When baby Landon was fourteen days old, in a four-page letter to Dr. Hilgard regarding aspects of the soil samples in the Cotton Culture Investigation report, Smith concluded with: "I have had to jump up and walk about with the baby every few minutes so that my breaks in continuity of thought may be laid down against him. He usually sleeps all day, but as the nurse is taking Sunday, of course he is making the most of it and has not slept at all."

The geological survey supplied Smith with an outdated version of LaTourette's map of Alabama, but a map more useful to him came from Colonel Daniel H. Cram of the L&N Railroad. (Smith mentioned it in his 1878 report to Governor Cobb.) The original hung on Cram's office wall in Birmingham, a large, newer edition of LaTourette that Cram had cut into one foot squares, mounted on muslin to make it foldable, and embellished it with his many notes and corrections. In April 1881 when Smith read Cram's death notice in the newspaper, he immediately sent off letters to men he thought would know, inquiring about the disposition of Cram's map: ". . . [T]his particular map with Mr. Cram's notes will be of the greatest service to me as State Geologist, and as I shall some day be obliged to get up a large map of the state for the Geological Survey, these notes ought to come to this office where they will be preserved and utilized." He was willing to pay for it, Smith wrote, "as much as anyone else, and perhaps a little more." On April 21, the map became the property of the Geological Survey; whether it was donated or Smith himself bought it is not known.

Smith had received no answer to the letter he wrote Postmaster Hansberger the previous October. He wrote again on April 8, 1881, repeating his request for specimens of the croton's flower at various stages of development. He warned Hansberger that the flowers were easily overlooked "but you will see them at the base of the leaves between leaf and stem." The same date, he wrote Dr. A. W. Chapman, "With your Southern Flora in my hands, I have had a great deal of pleasure in botanizing in Miss., Ala., and Fla., and would have gone to Apalachicola [Florida] last summer, but I was told you were not at home."

As soon as Smith finished his Florida report, daily attacks of fever and headaches forced him to take a "gigantic loaf," but he continued his regimen of letter writing. He thanked Dr. Hilgard, at Berkeley since 1874, for recommending him for the geologist position with a Yucatan expedition, an ambition

Smith cherished. Hilgard held the positions of professor of agriculture and director of the Agricultural Experiment Station, the first in the nation, at the University of California. Smith filled Hilgard in on his upcoming ten-day trip down the Alabama-Georgia line with a Georgia state geologist "in order to be able to mark out more understandingly the divisions to correspond with his." He confided how much he yearned for a calligraph, an early model typewriter, to help him with his reports, but he did not have the $60 the machine cost. ". . . I find that another baby just exactly takes every additional cent that I can make. I shall have to go into coal mining or some other business." He added that baby Landon "is a fine fat boy of great promise." He enclosed two programs of recent concerts given by the orchestra.

In a June 10 letter to Dr. Chapman, who had asked for information on the croton for his manuscript, Smith wrote that all of his material had been sent to Dr. Mohr. However he described the plant in detail: "It makes almost impenetrable 'privy thickets' as they are called, and the peculiar aroma of the leaves as you feel your way through, is exceedingly characteristic . . . Some of the bushes are ten feet high and the stems 2-3 inches thick. It grows low and spreading." Smith later wrote Dr. Mohr: "The seed pod is something like that of the castor oil plant." By now Smith had several plants growing well in pots.

On July 12, Smith asked Dr. Hilgard for news of the proposed Yucatan expedition. "I should like above all things to take part in this trip," he wrote, but it failed to materialize for him. Also in July, Daniel Langdon began working under Smith's supervision as a fossil collector hired for two months by Truman Aldrich at a salary of $30 a month plus expenses.

Aldrich had come to Alabama from New York State in 1871, the same year Smith began teaching at the University of Alabama. A graduate of Rensselaer Polytechnic Institute, Aldrich had, by 1881, left the Pratt Coal and Coke Company and founded the Cahaba Coal Mining Company at Blocton in northern Bibb County. Smith, who was very choosy about who accompanied him on his geology excursions, often included Aldrich. He assisted Smith's work in many ways. According to Charles Adams in Blocton, Smith termed Aldrich "one of the . . . most eminent tertiary paleontologists of all time."

By mid-July, Smith had the longed-for calligraph, and he was "doing nothing night and day" but printing out the Alabama cotton culture report with it. He soon discovered the machine's shortcomings—it had no distinction between upper- and lower-case letters. His was the tedious chore of going through the manuscript and underlining every letter that needed capitalization. He wrote the company asking to exchange his machine for the more convenient model. Privately he felt grateful that the census happened only once in a decade.

At this time Henry McCalley asked Smith to recommend him for a position with a college in South Carolina. He knew only that the school was Baptist. In a July 23 letter, Smith assured McCalley that he would see "some of the Baptist Brethern" in town that evening and would find out more from them. He added, "I hope you will get the place you want though I should be sorry to see you leave here . . . for my own sake." He went on to thank McCalley and Colonel Huckabee for sending specimens of the new plant that had now been officially named *Croton alabamensis*. Botanists considered it "a splendid addition to our Flora," Smith wrote. Its nearest relative was native to South America.

The croton was indeed a "splendid addition" to known plants. Though the March blossoms were insignificant, as Smith warned Hansberger, all summer the wind blowing through the leaves stirred up the "peculiar" aroma Smith spoke of. Others likened the fragrance to "freshly cut ripe apples" or "sweet shrubs at peak blossom." Then in fall the foliage turned a pleasing orange-red, with a silver underlining revealed by the breezes.

Smith made haste to let McCalley know that the South Carolina college required a qualification that Episcopalian McCalley lacked, that of being a Baptist. "I am sorry for your sake that it has turned out so," he wrote, "still I don't think a denominational school would be a very desirable place for any one who could not run a Sunday School." He also told McCalley that Aldrich and the conchologist he was sponsoring, Dr. R. E. Call of Iowa, had been in Tuscaloosa for several days. Smith arranged for Dr. Call, a collector of fresh-water shells, to meet McCalley in north Alabama later in the summer and for them to travel together for a while.

In an August 10 letter to Thomas W. Palmer in Snow Hill, Smith wrote how very glad the family was that Palmer was coming to visit them, especially eight-year-old Eugene Jr., who was "making great plans to be carried out with the aid of his friend Palmer." Smith also asked for detailed information about the geological layout of the country surrounding Snow Hill: "When I passed over that part of my road, I was in trouble about heavy roads, etc., and my notes were not full, and besides I did not then have time to stop and make the proper examination . . . An early reply to this will be very acceptable, and yet I do not wish to hurry you, as I shall start to Cincinnati next Monday [to attend the meeting of the American Association for the Advancement of Science], to be gone about a week."

Smith wrote another recommendation for McCalley on August 15, this time to the University of Mississippi to fill the chair of geology and natural history. For the last three years, McCalley had served as Smith's assistant in geology and chemistry, and for three summers they had worked together in the field. Smith wrote that "he is a perfect gentleman, as I know from intimate association with him both in camp, and at the University, places in which, as you know, all the latent meanness of a man is sure to crop out. Mr. McCalley is no part of a braggist, and what he proposes to know, you may be sure of his knowing . . . Not withstanding his slender pay here, he has done more actual hard work, than anyone at the University. The field work in North Ala. on which he is now engaged, he is doing gratis, and will do it well." Smith perhaps suspected that McCalley felt discouraged about his lack of advancement in his profession. Next day he wrote McCalley giving him a pep talk and listing all the ways he was well qualified for the Mississippi position. However, this opportunity did not develop for Henry McCalley, and he remained at the University of Alabama for the remainder of his distinguished career.

Smith's displeasure with the calligraph increased. On August 17, he wrote Dr. Hilgard, "I don't see why the Calligraph man did not put in a semi-colon, it would be of more service than the colon, which in general makes too much of a stop."

Julia Smith had turned four on August 7. Her father wrote a friend to thank him for ". . . the little present to Julia. She is now in Prattville, whither I am going in a day or two."

All year long, besides teaching his classes and writing his special reports, Smith tested rocks sent in from the four corners of the state and wrote each sender an analysis; he advised prospective students what they could do to prepare for his courses, or if they had to drop out of school for a while, how to keep up their studies. He exchanged scientific specimens and reports with professors of other universities; he wrote letters to editors and to office holders on public matters; and he critiqued colleagues' papers that he believed contained errors. He ordered books for the children and underclothes for them and for his wife. While "vacationing" in Prattville he continued his prolific correspondence, particularly to knowledgeable persons in Alabama and Florida for more information to use in his census report.

However, he now received a letter requiring a kind of answer he had never written before. Mr. Love, acting coroner of Statesville in Autauga County, requested analysis of "a portion of a dead body (part of a stomach and part of its contents)" that he had taken to the university and left with McCalley. He stressed the importance of the case and asked Smith to give the matter his personal attention. In his reply on September 10, Smith assured Love that he would take every care possible in proceeding with the analysis. He explained he would not return to Tuscaloosa until school opened on September 20, then allowing time for the analysis, he could perhaps have the results by October 10.

Already on Smith's bookshelves were what he called "the ordinary works on analysis," but now he ordered several others that not only gave "the full direction for qualitative and quantitative estimations of strychnine, arsenic, etc., but also statements as to the time after death, that organic poisons may be detected by analysis of the stomach contents."

As planned, Smith returned to Tuscaloosa when the university opened in late September with about 125 students enrolled. Jennie and the children remained in Prattville waiting for the return of Eugene's parents from a summer-long tour of the North so they could see their grandchildren.

Meantime Smith learned that the stomach to be analyzed had belonged to T. C. Taylor, an elderly and respected resident of Autauga County. A few days after his burial an affidavit sworn to by the physician, Dr. Gee, and the dead man's son claimed that the deceased "had been foully dealt with by some unknown person." The death symptoms suggested strychnine poisoning to the doctor. The coroner requested additional tests for arsenic and corrosive sublimate.

A dramatic account in Prattville's *Southern Signal* described the meeting of the jury at the Pea Ridge graveyard to oversee the exhumation of Taylor's body. Dr. Gee then removed the stomach and sealed it in the jar for delivery to Tuscaloosa. McCalley alone had access to the "odorous" mass though he frequently consulted with Smith.

Smith home, Tuscaloosa. The woman is possibly Jennie Smith, sitting with the Smiths's only daughter, Julia.

While McCalley carried out the complicated testing process, Smith had the pleasure of hosting a visit from Dr. Mohr—their first meeting after a long association. In October, when the report on Florida was at last completed, Smith sent it by registered mail to Dr. Hilgard, a risky venture since he had no other copy.

Jennie, finally on the way home by train, suddenly decided in Birmingham to continue to Nashville, where her parents had lived since 1875 when Dr. Garland was made chancellor of Vanderbilt University.

On October 30, Smith wrote Truman Aldrich that Langdon, whom Aldrich continued to fund, had spent the summer in Clarke County searching for zeuglodon bones and other fossils without the success Smith had hoped for. However, he had found "enough to make a great many exchanges . . . I shall request duplicate exchanges one for you and one for me."

In November, he and "some of the boys" found time to set up a telegraph line connecting Smith's home and his office with the asylum, the depot, and McCalley's room. McCalley also finished his final report on the Taylor case. It satisfactorily cleared the unknown person who was under suspicion as the tests showed nothing but "small quantities of digitalin and . . . still smaller quantities quinine and chinchonine" that Dr. Gee testified he had prescribed for the patient.

On Christmas Day 1881, Smith wrote Henry DeBardeleben asking for a gift close to his heart: "Do you remember some three or four years ago, when you and Mr. Aldrich were down here, and went over . . . my Chemical laboratory with me?

"If so you may remember my saying that I hoped when you and he had made a fortune out of the Pratt Mines, you would help me set up a first class Laboratory and Museum. You remarked that you would be the very men to do it, if your enterprise were successful.

"I foresaw then that a fortune would be made out of the Pratt Mines, and have been all the time in hopes that you were in earnest in what you said then.

"Now that the Pratt Mines have been sold for such a fortune, I thought I would write and find out whether you were still in favor of carrying out the plans then proposed.

"This has been the plan in my mind, ever since the time mentioned; viz. to put up a substantial brick building, arranged . . . in one part as a museum for depositing the collection illustrating the geology and mineral resources of Alabama, and also the other natural resources, such as timbers, woods, etc. This could easily be made the finest collection in the whole South.

"The other part of the building to be fitted up as a laboratory, for carrying out the analysis of ores, etc.

"At present time, the rooms in the University building which are devoted to the purpose above named, are filled to overflowing with the specimens. What had been collected in the last few years, lie still packed away in boxes, for lack of space. The Laboratory as you saw when here, is nothing but a cellar, and though I have lately been able to put in most of the necessary appliances, it can never be made anything more than a make shift; because the rooms are totally unsuitable. The building for a laboratory and museum ought to be planned and put up with special reference to its use as a Museum and Laboratory.

"If you should favor the plan, a suitable building could be erected on the grounds of the University, to be called the DeBardeleben Museum and Laboratory, and devoted to the permanent exhibition of the specimens illustrating Alabama Geology and mineral and other Natural resources, and to the uses of a laboratory, where the young of the state could be instructed in practical Chemical analysis, etc. Your name would in this way be always hereafter associated with the instruction given to future generations of young men regarding the natural resources of the State of Alabama, as you have, more than any one else, contributed to the development of these resources.

"A building could, I think, be put up for $25,000 which would not only be an ornament to the state, but an everlasting witness to your liberality.

"If you should favor at all, the plan proposed, let me send you some plans of what would be a suitable arrangement of the interior of the building; the outer details would of course, have to be left to your taste, or to the architect.

"At any rate, whether you should agree to the proposition or not—please let me know if you receive this letter, as I am not certain of your address." DeBardeleben's reply, if any, is not available, but he was not known as a generous man.

Three days after Christmas, Smith wrote that he had accomplished a great deal of work on his reports over the holidays. "One day I wrote up four counties and had a large company to dinner besides; but they were all ladies and did not interrupt me." That same day, he wrote the calligraph man for the third time in reference to exchanging his machine; as yet he had received no answer.

1882

Choctaw, Clarke, Marengo, Monroe, Sumter and Washington Counties

The New Year began with Smith "about strapped as regards Geological Survey money and don't know that I will have enough to pay for the work." At the end of January, he wrote Professor R. P. Whitfield:

"You know that lack of funds obliges me to make all sorts of detours in getting anything done in the cheapest way."

He did not take to the field until August. On the sixteenth, Smith, McCalley, and Langdon, with Anderson driving, explored south of Livingston, where Smith's interest centered on "a dark blue, nearly black laminated clay" that he found in the flatwoods and the railroad cuts. Wells dug in this clay produced "water which is milky in appearance and tastes as if a dirty gun barrel had been washed in it."

Three days later, the group left Sumter County, crossing the Tombigbee River to Marengo County. "The ferry is the worst yet seen in Alabama," Smith judged. They had to climb straight up about twenty feet, and even with an empty wagon and "all of us at the wheels, we came within an ace of going back into the river."

In the last week of August, heavy rains set in "which beat through the tent making us all thoroughly uncomfortable. . . . It was impossible to cook in camp as the water put out our fires by running in below and pouring down above." In the night with rain still pounding on the tent, Smith felt the forewarning of a chill. Next morning the expedition found an available house where Smith went to bed with a full-blown case of chills and fever. While he continued disabled, McCalley and Anderson took the wagon to Woods Bluff for collecting, and Langdon rode a considerable distance to Bladon Springs for the mail. Langdon returned bringing news of both McCalley's and Anderson's illness at Woods Bluff. Smith sent Langdon to fetch the two sick men, and all of them were taken in by Captain Stimpson. The heavy rains continued, but at Hoboken the expedition was fortunate to get "a good new house." Langdon had to retrace his steps searching for a catalogue that he lost the day he brought the mail from Bladon Springs.

The geologists left Hoboken on September 1 under a cloudless sky, but within two miles they were into heavy rain that lasted intermittently all the way. They "came either through rain or just after it, getting full benefit of suffocatingly hot steaming weather and deep boggy clays wet up by recent rains."

In Choctaw Corner, they took shelter in a vacant office provided by Dr. Mobley. The rain continued to descend "in torrents." When the weather finally cleared Langdon left for Claiborne, while McCalley and Smith went off to Tallahatta Springs. On the way to St. Stephens on September 6, Smith and

McCalley viewed the freshwater shell middens in the Indian settlement of Turkey Town on Mr. Thornton's plantation. Smith collected from Thornton "a perforated stone," and "saw in his yard a large block of fossil palm wood" that had been found close by.

Approaching St. Stephens, once the territorial capital of Alabama, heavy rains caught them again. "McCalley who had no umbrella got thoroughly wet, but I got wet only on the knees." Smith wrote later. They traveled through pinewoods being tapped for turpentine, and passed a turpentine distillery just west of St. Stephens.

On September 8, they rode to the St. Stephens Bluff, which rose one hundred feet perpendicular from the river. It was "made up . . . of the soft white limestone such as is used in the construction of the chimneys of this section." From there Smith took a boat to [Old] Washington, his Autauga County birthplace, then visited his parental family in Prattville. By September 20, 1882, he returned to Tuscaloosa.

1883

Warrior, Tombigbee, and Alabama Rivers; Clarke, Dallas, Jefferson, Marengo, Monroe, Wayne, and Wilcox Counties

The university's Board of Trustees granted Professor Smith leave from his teaching obligations for the fall 1883 term, allowing him to fulfill his duties as state geologist. On August 15, he and Judge Lawrence Johnson of the United States Geological Survey, boarded a hired steamer, the *Ella D.*, at Tuscaloosa with the intention of studying the Cretaceous and Tertiary strata of the Warrior, Tombigbee, and Alabama rivers.

They did not get off to a promising start. "We tied up for the night at Bozman's plantation, and during the night there came on a heavy rain, and as our roof leaked like a sieve, we had a very uncomfortable time of it and very great difficulty in keeping dry." After breakfast on August 16, they "started down the very crooked shoals and got stuck in the sand at the foot of the last turn. Here we remained till after night fall, when we succeeded in getting clear: tied up then till day."

They had not yet cleared Tuscaloosa County when they "discovered that the 'Bustle' on one side [of the *Ella D.*] was torn loose and jammed a rudder and spent the rest of the day trying to remedy it." Smith did not let these mishaps deter him from making notes on the clays in the embankments, and the springs and artesian wells along the way.

The boat stopped near Logan's Bluff on the eighteenth "to take on wood from a drift pile." Smith used the time to collect six "representative samples of the second Bottom Deposits [that] are seen with such (apparent) uniformity of composition, structure and appearance on all the rivers in Alabama."

The next day, Smith left the *Ella D.* tied up at Finch's Ferry and went back to Tuscaloosa just long enough to fetch things he had forgotten and "a workman to patch the boiler." On the twenty-first, above Melton's Bluff the *Ella D.* hung up on a log. They "stayed there till after sunset when we drifted down to the head of the bluff and tied up for the night. Boiler leaking still like a small branch. Engineer threatened with fever."

Near Owen's Bluff the following day, a swift current caught the *Ella D.*, tearing off "two or three of the wheel arms and two buckets." They narrowly missed "getting the stern of the boat torn away," Smith wrote, "but fortunately lodged across a log where with the assistance of a stern line, we held up and repaired the damage and were able to get to Hines Bluff and get on several cords of wood before night."

On the ninth day after leaving Tuscaloosa, they reached Demopolis about 4 p.m. where they picked up mail and bought a few necessities. Dr. James Whitfield, a classmate whom Smith had not seen since undergraduate days, came aboard for a visit.

On August 24, the *Ella D.* paused to visit the *Big B* with Captain James O. Abbott aboard. The captain gave Smith a steatite bowl that had been brought up from the river floor not far from Rembert's Landing. When the geologists passed the *Ruth* tied up for the night, they stopped to exchange news and make inquiries. The *Ella D.* spent the night at Stevens Landing. Before breakfast on the twenty-fifth, they loaded two cords of wood, and while they were eating, the *Ruth* passed on the way downriver.

They tied up at Woods Bluff and spent most of the day collecting. They welcomed Captain Stimpson aboard to spend the day and night with them. It was he who had taken the expedition into his home the year before during a miserable time of sickness and inclement weather. On the twenty-seventh, they were once again at the Indian settlement they visited the previous year on L. S. Thornton's place near the mouth of Oaktuppah where "shells and pottery fragments are . . . abundant."

At Bladon's Landing, Smith wrote that "here again is presented the puzzle of the past three or four years, where the two opposite sides of the river show such entirely different geology . . . . I can't begin to understand it all yet but hope to do so before the day is over." On the twenty-eighth near St. Stephens, they stopped to collect a number of fossils from the shore. True to his intuition Smith came to an understanding of the "curious anomaly" of the rocks on the Choctaw and Washington counties side of the river differing from the Clarke County side, entering the diagrams in his journal accompanied by a detailed explanation.

The night of the twenty-eighth they tied up at McIntosh's Landing where at daylight next morning the *Ruth* passed on the way upriver. Her captain relayed to the *Ella D.* a message from the inspector in Mobile—if they went up the Alabama River as planned, without a pilot, their captain risked a $150 fine. That news necessitated the *Ella D.* continuing to Mobile to pick up a pilot.

Smith, as usual, turned this setback to advantage. When they arrived in Mobile on Wednesday, August 29, about six p.m., he went into the city "where I met with Dr. Mohr and took tea with him."

Next day the *Ella D.*, with the required pilot aboard, headed upriver. "At fifteen miles above Mobile we landed at one of the shell islands and got a number of Gnathodons of which the island is made up." August 31 "was devoid of interest" as they proceeded up the wide, shallow lower river with its large bends. The boat "had to keep along pretty closely to the outside of the bends, making our way considerably longer. Yet we got up to Choctaw Bluff before dark and tied up there for the night." Next morning Smith examined the bluff and entered sketches of its sections in his journal.

Later at Gainestown he explored the abandoned quarries, coming away convinced that the beds underlying the limestone contained zeuglodon fossils. "This matter deserves a closer examination."

Smith spent the morning of September 2 collecting "some of the rarer forms" at Claiborne, but in the afternoon he suffered a disabling headache. At the Falling Spring next day, he had "several barrels of the fossiliferous sands" brought aboard, and filled numerous pages in his journal with notes and diagrams of the section of the strata he found exposed on this bluff. At

Bell's Landing on the fourth, he collected and sketched gigantic specimens of fossil shells.

Trouble again on the fifth when the *Ella D.* ran aground near Clifton, causing the pump to clog. They had to tie up "and draw the fires till the valves could be cleaned and the pump put in order." On September 6, Smith left the *Ella D.* at Prairie Bluff, and boarded the *Mary* bound for Selma. He arrived there at midnight and eventually returned to Tuscaloosa.

He was not home long, however, before setting forth with the wagon on October 6 for Fayette and the coalfields. On the tenth, he visited McCalley's camp near the mouth of Lost Creek and stayed overnight. About two miles from Village Springs on October 23, one of the horses came down with colic, and T. S. Woods invited them in for a day and a night. The last week of October, between coalmine tours, Smith attended the opera in Birmingham several times before returning with the wagon to Tuscaloosa.

In his report of this trip to Governor Thomas Seay, which was included in the Geological Survey Report of 1884–88, Smith stressed again the difficulty and expense of assessing the coal fields. "Generally the outcrop of a coal seam is hidden by dirt and soils and debris of all kinds, so that to expose it to view it must be uncovered, and, unlike most other valuable minerals, a coal seam often leaves upon the surface no sign of its existence below . . . . To acquire a thorough knowledge of a coal field is one of the most difficult problems which can be presented to a geologist or to a miner."

Now by the end of the crucial first decade of the revived Alabama Geological Survey, Smith had visited every county in the state. A glance at the subjects of Smith's publications during this time, as compiled by Dr. Walter B. Jones, underscores his single-minded determination to ferret out Alabama's natural resources and make the information available to the public: a geological and agricultural survey of Alabama, 1873; the crystalline rocks, 1874; the Paleozoic rocks, 1876; the Paleozoic valleys and Coosa coal field, 1876; outline of Alabama geology, 1878; the Tennessee Valley and Warrior coal field, 1879; the iron ores of Alabama, 1879; the survey of the Black Warrior River from Tuscaloosa to the Sipsey Fork, 1880; an in-depth description of the northwestern quarter of the state, 1881; an in-depth description of the agricultural features of Alabama, 1883; the iron ores of Alabama and their geological relations, 1883; to name but a few.

Industrial leaders such as Henry DeBardeleben and Truman Aldrich read Smith's bulletins, reports, and articles in scientific journals. They consulted him personally. Ordinary citizens as well came to depend on his guidance in their enterprises and to trust more fully in his judgments. One of his most valuable assets was his ability to attract men of exceptional ability to work with him for no pay.

In these ten years, Smith had of necessity become skilled in the political process. He remained continually alert to events in Montgomery and Washington, protecting the survey's existence and its integrity, lobbying for more generous provisions to do the job that Alabama needed.

At the end of the first decade the state legislature responded to the accomplishments of the Alabama Geological Survey by raising Smith's annual allowance to $5,000. The new allocation for the first time allowed Smith to pay salaries to his assistants: $175 to McCalley; $150 to A. M. Gibson and also $150 to Joseph Squire, who had more than thirty years experience as geologist and mining engineer.

And at last Smith's dream wagon became a reality. On March 14, 1883, he wrote the Studebaker Brothers Manufacturing Company describing the wagon he wanted. They replied advising him that a white canvas top "to be well painted with three coats of red lead, this would be better than the board top of which you speak . . . . It would take us about thirty days to get it out. It usually takes 60 days to get out a job of this kind, but we have one partly commenced that would fill the bill." Smith replied on April 6. Studebaker countered with more advice on April 17. They added, "We did not think about your sleeping in the wagon; but enclose you photograph of a wagon which we build for Salt Lake and Colorado trade. The backs of these seats are on hinges and drop down to make a bed 6 foot long, and make a good bed at that, as they have cushions stuffed and are trimmed with oil leather, gotten up in first class style." The company noted what these seats and the extras Smith wanted would cost and added, "We would be glad to hear from you, and please say if you want us to build the wagon or not."

Dodd Sellers described in detail the efficient arrangement of the "special order" Studebaker, which still exists, in the January 1983 issue of *Turning Wheels*: "The wagon, which cost the State $260 ($100 down and balance in 90 days) plus $21 shipping charges, was constructed of hickory and oak and had iron tires . . . . An apron on the back of the wagon carried camp equipment, folding table and chairs, dutch oven, tent and cots. The backs of both seats were hinged and when let down, converted to a bed. There is a storage drawer under each seat. Tarp flaps on both sides and in back could be let down to provide privacy or to keep out rain." The color scheme of the wagon, which arrived in Tuscaloosa by rail, combined dark green and yellow, and the upholstery was leather. This carefully crafted beauty served the Geological Survey team well until a Model T Ford car replaced it in 1904.

2

The Second Decade, 1884–93

1884

Bibb, Fayette, Greene, Perry, and Walker Counties

In March of 1884, Smith wrote Truman Aldrich that he wanted him to see the "very handsome" set of shells Smith had assembled for the Birmingham Exposition. Smith, however, had a more serious request. He asked Aldrich to exert influence on his Iowa friends in Congress to get a bill passed without further restrictions granting public lands to the State of Alabama in recompense for the intentional burning of the University of Alabama by Federal troops on April 4, 1865. Brigadier General John T. Croxton had ordered the burning although there was no Confederate armed resistance in Tuscaloosa. The bill provided that the governor of Alabama appoint agents to choose the lands, and after the U.S. secretary of the interior approved the choices, the lands were to be sold to benefit the university. The bill had already passed the senate, Smith told Aldrich. He added, "I don't see any way of getting a museum building or a laboratory unless from this source and you know our needs in this direction."

Smith busied himself assembling a hundred or so wide-mouth bottles "the size and shape of the 4 oz. quinine bottles" and filling them with soil samples for the Birmingham Exposition. Langdon had finished mounting the Claiborne shell series and planned to send them, the coal samples and the coal plants before the exposition opened on April 22. McCalley was already in Birmingham setting up the displays.

The land bill, allowing Alabama 46,080 acres of public lands in reparation, passed the United States Congress on April 24, 1884. When Smith was tentatively offered an appointment to the commission that would choose the lands, he replied with alacrity that he would be "as much interested in selecting valuable lands, as anyone could be." The note he added summed up his guiding principle: "You may count upon my doing what I can towards advancing the interests of the University."

Earlier in the spring, William Spencer of Perry County had sent Smith a box of greensand. Tests showed the specimen contained a high percentage of phosphoric acid especially valuable to agriculture. As soon as possible, Smith visited the Spencer plantation near Hamburg. He found the ground covered with phosphate nodules; in just one spot he picked up two pounds. The next day he wrote Governor E. A. O'Neal: "I have the favor and pleasure of announcing the discovery of a bed of phosphates . . . A very large number of analyses will yet have to be made before even the approximate value of these phosphates can be estimated; but I have very little doubt but that they will form the basis of a great industry in the future, especially since this belt . . . probably extends across the whole width of the state."

To his father, Eugene described the phosphate nodules as having "very irregular and curious shapes," colored from "light blue to bluish black," and having "a peculiar odor" when rubbed together.

The phosphates discovery caused a great deal of excitement not only among reporters and the public, but also among Smith's colleagues. Judge E. C. Betts telegraphed Smith asking for the exact location of the find, so he could send Professor Stubbs from Auburn to investigate the beds. Smith was wary. He replied: "Whilst of course I have no objections to giving you the fullest information as to the localities occurrence, I should be very sorry if any one should take this discovery and start in to working it up for publication in advance of my own publication . . . I shall be able to give my results to the public in the course of a month or two, and as a scientific reputation is all that I can expect from the work, and as it is of very great importance to me, I am anxious to get the first fruits of the discovery, which I could not if you should put a force into the field to work on the subject ahead of me."

Smith made haste to send Langdon to investigate the phosphate find thoroughly. On May 4, Langdon wired Smith in Tuscaloosa that notwithstanding his letter to Judge Betts, Professor Stubbs had appeared at the site. Smith replied: "I do not think anything can be gained by my coming down and besides Prof. Stubbs may not have in mind the intention of publishing his searches . . . I hope you and he were able to agree upon general terms, and that he will be . . . fair and square, but if he should not, then we will have to let him take his course." Smith quickly dispatched further details of his discovery to the newspapers in Chattanooga, Montgomery, Selma, and Mobile.

In the midst of the phosphate turmoil, Smith took his place as second violinist in Professor Chisholm's orchestra at the laying of the memorial stone for Clark Hall on May 5. The stone contained, among other mementoes, an 1821 silver dime taken from the cornerstone of old Franklin Hall, burned by Federal troops that fateful April.

Smith received a letter from Professor Stubbs dated May 16 that he answered immediately: "I am glad of the opportunity which it gives me of speaking freely on a matter of personal concern to us both . . . I may say frankly that I should never have thought of putting even the slightest obstacle in your way but for the fact I have noticed . . . in your published reports and other documents you do not generally give to other workers in the same field with yourself . . . full credit for what they have done." Smith pointed out two instances where Stubbs had made extensive use of Smith's work and given him no credit—his manuscript on the Coosa Valley and his report on the Warrior Coal Basin exploration. Of the latter, Stubbs had written Smith, "I have followed your two reports closely . . . and in making my report shall draw largely on yours." Yet in his manuscript, Stubbs not only gave no credit to Smith but wrote: "Hitherto, but little was known of this basin, being a veritable terra incognito, except an occasional exaggerated account . . . of coal beds of enormous thickness and wonderful extent."

Smith also wrote Judge Betts at this time repeating his objections to Stubbs moving into the phosphate fields ahead of him. He offered to go into detail as to his reasons, but at this time wrote only that Stubbs probably did not deliberately intend "to create a wrong impression . . . His losing sight of the claims of others where he is personally concerned, is I think, in great part due to his temperament and disposition which as you know is to make the most of what he had done himself. In the matter of scientific work, I have also the desire to get credit for every thing I do, for scientific work is what I have devoted my whole time to."

Truman Aldrich went with Smith to Hamburg to evaluate the phosphates in the two pits that Langdon had dug. They concluded that there were not enough nodules to support a profitable business venture. Smith also assessed a phosphate discovery in Eutaw, but he found the nodules less abundant than at Hamburg. He remained convinced, however, that in the 150-mile range of the phosphates "really important accumulations" would be found. The market was ready. "The manufacturers of fertilizers from all parts of the country are sending now for information," Smith wrote. He had already allocated most of the annual appropriation from the legislature to the study of the coalfields. He regretted he had only a pittance left to spend on the phosphates investigation—he yearned for ten thousand dollars.

By August, Smith, now an official member of the three-man land commission, left Tuscaloosa by way of the Crabb Road ("without doubt a worse road than the Watermelon") to begin his appraisals. While in Walker County, he made a round trip by train from Corona to Berry Station: "This part of the road passes through a beautiful country and the way is over seven high trestles, one 97 feet, through deep cuts and over high embankments. I don't suppose there is any road in Ala. which has more imposing scenery."

In the course of the land search, Smith devised a plan whereby the Geological Survey could acquire a much-needed diamond drill. He proposed to Governor E. A. O'Neal that the state buy the drill for the survey, then rent it to the federal land commission to aid their choice of the most valuable lands for the university. The rent from the commission would help the state pay for the drill.

By September 1884, 209 cadets had enrolled in the university at the cost of $76.50 for a half term. Smith, along with other professors, continued his efforts to up build the library's resources. What had been a thirty thousand-volume collection, now—nineteen years after the burning of the university—consisted of six thousand books.

Smith also kept alive his dream of a museum-laboratory. When he went to Philadelphia for the meeting of the American Association for Advancement of Science, he admired along the Shenandoah Valley route a number of public buildings designed by architect George T. Pearson of Philadelphia. In October, he approached Pearson with his rough sketch and his $25,000 limitation hoping to arrange for a "pretty and taking design" to show the university's Board of Trustees. However, Pearson required $25 for a plan, $25 that would have to come out of Smith's empty pockets.

At this time, Truman Aldrich pressured the Geological Survey to make a trip down the Cahaba River, but Langdon, who would be the one to do it, was occupied with the phosphates. Smith wrote Aldrich on October 21 that he would hold the phosphates in abeyance allowing Langdon to make the trip if Aldrich thought the Cahaba could not wait. This was a notable concession on Smith's part, because he was convinced that the phosphates were "bound to work a revolution in the farming interests of Ala . . . I think the sooner the deposits are thoroughly examined and tested, the better."

Smith wrote Governor O'Neal on October 26, "I have thought that perhaps . . . the recent discovery of phosphates in Ala., might be of sufficient importance to merit a place in your message to the Legislature. If you should think so too I should be much gratified if you would make mention of it—and to save trouble I have enclosed an outline of the subject, the substance of which might, if you thought proper, be incorporated in your message.

"The subject is one in which every farmer is bound to be most deeply interested, and the great multitude of letters and specimens which I have received since the first announcement of the discovery—has shown me that this matter is likely to be of more importance to the welfare of Ala., than all the coal and iron, and it will certainly deeply concern a much larger proportion of the population than coal or iron does." He went on to praise the legislature for the recent increase in funding for the Geological Survey, and recounted how he worked for the first ten years without financial reimbursement. "All my hopes and ambitions are centered in this Survey, and you cannot therefore wonder, that I should feel pride in the consciousness that the works of the Survey will certainly . . . prove of lasting benefit to a lage proportion of our population and if we take a business view of it, will pay back to the state with interest, every cent that has been or will be expended upon it."

By October 30, Langdon had started on his way down the Cahaba River from Centreville searching for sub-Cretaceous fossils despite Smith's continuing hints to Aldrich that the trip should be postponed until rains relieved the drought conditions. On November 1, Smith acknowledged to Judge Johnson the receipt of his manuscript of their *Ella D.* river trip and expressed his aim to get at his part of their joint paper soon "but don't know, as I have a perfect stack of lectures—not withstanding I write on them an hour or so every night." (This report was not published until 1888.)

The coveted diamond hand drill had arrived, and Smith sent William A.

Gould upcountry to test it in the field. Gould returned much pleased. He found it cheaper by half to operate than a steam drill, yet it performed almost as well. When Smith reported this to Colonel Hargrove, another member of the land commission and a state legislator, he also confided how disappointed he was that Governor O'Neal failed to mention the Geological Survey in his message to the legislature. "Every other state Institution is mentioned," Smith lamented, even though some of them did not even send in a report as he had done.

Governor Edward A. O'Neal (1882–86) reduced taxes, yet left a surplus of over $300,000 in the state treasury. He established the Board of Convict Inspectors and improved unsanitary conditions, insufficient food, and poor health care in the prisons. He founded several normal schools, including Livingston and Jacksonville; and established the State Department of Agriculture among other accomplishments. Considering his record, O'Neal's indifference to Smith's informative and affecting letter was hard to understand.

On November 18, Smith wrote Aldrich that Langdon had sent word that he would be returning any day, bringing a scant and undistinguished collection of fossils from the Cahaba.

1885

CLARKE, CHOCTAW, HALE, MARENGO, MOBILE, MONROE, PERRY, SUMTER, AND WASHINGTON COUNTIES

The field trip that left Tuscaloosa in July 1885 aimed to fill in the gaps that were left in the river trip investigation made by Professor Smith and Judge Johnson aboard the *Ella D.* in 1883. Smith's field notes would now have an added dimension for he had bought a camera and applied himself to learning how to use it to best advantage. South of Carthage (Moundville) in Hale County, he photographed the yellow sands with pebbles and the red loam, which was a deeper red than the hills of Tuscaloosa. From there the expedition went to Dayton, Greensboro, and Newbern, all the way dogged by heavy rains. At Newbern, they took refuge in Dr. Higgins's hotel, then on through Uniontown and four miles beyond where they put up with John A. Jones. "It rained heavily so that we were very glad to have the shelter of a house," Smith wrote.

After studying the numerous shells in a gully across the Chickasabogue,

Two photos taken by Smith on July 11, 1885. Top, a gully at Havana, Hale County. Bottom, the gully near Sanders Ferry Road, ten miles from Tuscaloosa, one of the deepest Smith, an authority on gullies, had ever seen. In the bottom flowed a "fine stream of clear cold water" treacherous with sandy clayey spots on its banks.

Smith sent the wagons on ahead while the collectors including Aldrich and Langdon went downriver in a skiff. At Midway Landing the large Nautilus shell they tried to extricate broke, but at W. T. Matthews' Landing on the left bank, they collected for several hours and felt satisfied with "a lot of fine shells." Smith commented on how the black clays near the mouth of Dixon's Creek had crumbled "setting free the beautiful fossils," and he made a very long list of them in his journal.

On July 19, they left Lower Peach Tree about 9 a.m. bound for Black's Bluff, eleven miles away, which they reached shortly after noon. The following day they spent a couple of hours examining the mile-long bluff below Gregg's Landing before moving on to Bell's Landing. Here they descended the bluff in the unrelenting July sun, "and a hotter place I have never yet had the fortune to see," Smith wrote later. At the foot of the bluff, Anderson awaited them with the wagon to load their specimens that now included the large shell, *Voluta newcombiana*, found in few other places besides Bell's Landing. Dr. Mohr met them at Perdue Hill and joined in an afternoon of shell collecting at the legendary Claiborne Bluff, called the "El dorado of Geologists" by Michael Tuomey.

Years later Smith wrote of Claiborne: "Half way down the bluff there is a bed of yellowish sand some seventeen feet in thickness and extending along the face of the bluff for at least a mile, in which are embedded those beautifully preserved fossil shells which have made the name Claiborne famous in all parts of the civilized world where museums have been established. This bed might . . . be better described as a bed of shells packed in sand, which merely fills in the spaces between them. The mass is only slightly coherent so that one may break down from the bluff a great lump of it, which may be crumbled up with the fingers, liberating the shells in perfect state of preservation and strong enough to be not too easily broken. While the markings in color of the original shells are generally obliterated, one finds now and then a specimen in which these markings are still to be clearly observed, notwithstanding the vast lapse of time since the shell held a living being.

". . . During the eighties, while the Alabama Geological Survey was engaged in working out the details of the tertiary formations of the State, large collections were made of the Claiborne material, and a number of new forms were discovered by Mr. T. H. Aldrich of Birmingham, Mr. Otto Meyer, and by Mr. D. W. Langdon of the Geological Survey . . . [So] far as is yet known the ferruginous sand with its beautiful fossils has not been identified beyond the limits of Alabama."

The night of July 25, 1885, the geologists camped at Gainestown where with great difficulty they got the wagons and mules on and off the ferry. Here Dr. Mohr left them after giving the limestone bluffs a thorough examination. Upon their approach to Jackson on July 28, Smith noticed "a good lot of agate, chalcedony, and other forms in the road, some of it was very pretty and might answer to polish." The end of July found them camped beyond Womack's Hill on Tallawampa Creek, Choctaw County, where Smith found clays so pure that he believed they could be used in the making of pottery.

Aldrich and Smith, leaving Langdon at base, took the buckboard scouting around and collecting through the canebrake to Wood's Bluff for an overnight stay. "Here we collected for several hours in the very hottest place in the whole world. Mr. Aldrich became over come with heat twice and had to go up on the bluff." Smith awakened next morning "with pains in every part of me" followed by fever later in the day. Even so, he went with Aldrich in a skiff downriver, but the notes in his journal for that day were furnished by Aldrich, as Smith was too sick to exert himself. "Paddled back to Wood's Bluff in the hot sun, with fever and headache," he wrote later, "and was glad enough to lie up on the bed while Mr. A. packed up his shells and harnessed the horses. As we started back there came up a very heavy rain which made the driving through the canebrake very disagreeable. I was tortured all the way with back ache and headache, and did not enjoy it much."

Back at camp they found that Langdon, sick with chills, had been taken in by Steve Noble, their nearest neighbor. Noble now insisted that the rest of the party spend the night with him "which we did to our great satisfaction as we were well treated," Smith recorded in his journal.

On August 12, Smith, Langdon, and Aldrich followed the Pushmatahaw Road from Intercourse into a forest of giant long leaf pines where "a great cloud and a threatening of rain and wind . . . made us stir ourselves to get out of the high pines." They reached their destination, Mr. Baskins's house between Tickabum and Yantley creeks, before the rain began. On August 18, they camped at Coffeeville. Next day they left the wagon at the five-mile post from Grove Hill and went to collect Claiborne shells, some of them of "extraordinary size," Smith wrote. "We filled several boxes with the shells and one with shells and sand just as they are found." In this area, they added zueglodon vertebrae to their collection, and on the twenty-first Smith took a "first rate" photograph of the bluff at St. Stephens, "the highest part upon which the old fort stands."

During this time they had to change camp several nights because they could locate no corn to feed the horses. Near Suggsville, after climbing a long red hill, Smith discovered a knob of the Buhrstone clays. "On the same hill and not fifty yards distant," he found the "horsebone" (orbitoidal) rock. "This close proximity of the two series of rocks which are separated by about 200 feet vertically, is difficult to explain."

A cadet from the war years at the university, A. R. McConnico, hosted the group the night of the twenty-eighth, and next day guided them in an exploration of the Flat Creek area. Langdon, feeling the chills coming on, started back to camp but got lost and did not straggle in until nearly an hour after

The survey crew's camp near Allenton, Wilcox County, September 10, 1885.

Colonel J. W. Lapsley's home near Vinton, Autauga County, September 16, 1885.

the rest of the party returned. "He got caught in three rains just as we did," Smith wrote, "and got wet through each time, all of which appeared to do his fever good. The quinine which he took kept off the chills and a good part of his fever and he is OK now."

From McConnico's, they took the Allenton Road for ten rugged miles over gray clays made heavier by the rains. "The wheels brought up hundreds of pounds of clay which was held firmly together by pine straw. It took us three hours to come six miles." When they finally gave up the struggle, they camped "in the road and put up our beds in a cotton house."

September 9–10, camping near Allenton in Mr. Jones's Gin Yard "in a very pretty grove," Smith noted that "these prairie soils must be extremely rich for the weeds, especially an Artemesia, grow to the height of ten feet or more. In a good deal of this . . . the rocks cover the soil almost completely and yet fine crops are produced." Smith examined some grayish white limestone at Palmers Mill that was "a dark gray of bluish color when fresh and wet and is then quite soft and easily cut so that it is extensively used about here for chimneys, as it hardens on exposure and resists heat. This rock gives a black loose soil which sustains in many places dense thickets of cedar, hence call Cedar Brake land." Later he noticed this rock also being used for hearths.

On September 15, Colonel J. W. Lapsley entertained the survey crew with dinner and a tour of geological places of interest. Next day after dinner all the guests including some ladies rode out two miles to a ravine to see an exposure of the Tuscaloosa formation. Everyone posed for Smith on the Iron Bridge over the Mulberry Creek at Vinton with the Tuscaloosa formation in the foreground. Smith seemed to savor the words as he named the beautiful colors of the rocks and soils: purple, red, laminated gray, yellow, pinkish, orange, white, blue, chocolate brown.

Less than a month after Smith's return to Tuscaloosa, his brother-in-law, forty-eight-year-old Burwell Boykin Lewis, president of the University of Alabama, died in the president's mansion at 2 p.m. on a Sunday. Married to

Colonel Lapsley entertained the survey crew and other guests in September 1885. On the sixteenth, everyone "including some ladies," rode two miles to a ravine to see an exposure of the Tuscaloosa formation. The party posed on the iron bridge over Mulberry Creek at Vinton.

Jennie's sister Rose, Lewis had seven daughters. The funeral, held in Clark Hall, was attended by Governor O'Neal, the Garlands from Vanderbilt University, most of the university trustees and an enormous crowd. The following May, the university orchestra (sometimes called Professor Chisholm's Orchestra) took part in a memorial concert honoring Dr. Lewis.

Eugene and Jennie's third son arrived in December 1885. They named him Truman Aldrich for Smith's valued friend.

1886

BUTLER, CHOCTAW, CLARKE, GREENE, MONROE, WASHINGTON, AND WILCOX COUNTIES

On Monday, July 12, Smith and Langdon took the wagon out of Tuscaloosa by way of the Sanders Ferry Road. Ten miles from town the road skirted an impressive gully, one of the deepest Smith had ever seen, remindful of the Western canyons pictured in geological publications. In the bottom flowed a "fine stream of clear cold water," treacherous with clayey sandy spots on its banks. During their exploration and mapping of the strata, Langdon inadvertently sank "right arm foremost nearly to his body" in the quicksand. Next day they examined other similar gullies and spent the night at W. S. Patton's farm where they saw the lotuses blooming in his millpond and admired his herd of Jersey cows and his carp pond.

By the sixteenth, they were taking tea in Eutaw. From there they explored between the Black Warrior and the Sipsey rivers and visited Mr. Anderson who, three years before, had spread on a field the greensand thrown from his well. The present crop on that field was still flourishing from that one application of greensand.

Smith continued to make use of the camera; he changed the photographic plates "by means of the bag and came very near melting down with the heat." He made many exposures in the area of Black Bluff, which he last visited in 1883. The spring 1885 earthquake had changed it drastically causing "the highest part of the bluff just below the landing . . . [to slide] off into the river . . . and . . . a very large area (several acres) some little distance back of the immediate river bluff sunk down a distance of ten feet or more." From the bluff, they took a skiff for Thompkinsville, sending the wagon on ahead. At

As Smith drove through the geologically-dull piney woods, he studied the flowers in bloom "in great abundance and beauty, as only the piney woods flowers can be." August 15, 1886.

Naheola on the Tombigbee at the end of July, they packed up their collections, arranged the skiff's return to the Bluff, and rejoined the wagon.

Along the way to St. Stephens on the Alabama River, they met William Lenoir who gave them many notes not only on soils, rocks, and outcroppings, but also corrections of the map and botanical lore. He told them that a tea brewed from elephant's foot (*Elephantopus carolinius*) would prevent the chills so common in this fever-ridden area.

The soil at St. Stephens, Smith wrote, "is capable of indefinite improvement and beautiful farms could be maintained here, especially by the free use of the Jackson Marl at the base of St. Stephens Bluff," which was rich in phosphorous. At Gainestown the surprising sight of "a beautiful house and well kept grounds belonging to Mrs. C. L. Jewett" so moved Smith that he took a photograph of it at three p.m. in sunlight, four seconds' exposure.

On August 18, they set up camp near Watson. Anderson, their driver, had come down with a fever, and Smith himself was so unwell he stayed in until late afternoon. Meanwhile, Langdon spent the day collecting at Falling Spring. The following day at Claiborne they boxed and shipped their specimens and that night found an old house to shelter them for Anderson's sake.

They spent August 22 socializing with visitors at McConnico's, Smith's wartime comrade, and next day made camp twelve miles south of Camden "in front of a house from which the occupants had gone leaving a puppy and a cat. They were probably visiting a sick neighbor and we tried for some time to call up someone here and at the next house below, but failing to get an answer we went into the yard and found a well, which was all we absolutely needed." As soon as they crossed into Wilcox County a steady rain began with the promise, Smith thought, of continuing all night so "we have the usual prospect before us of Wilcox limy clays." Upon reaching Camden, they put up at the hotel because of the weather.

A boat trip from Wilcox Landing on August 26 ended at Prairie Bluff in the rain where they had to carry all their "plunder up the hill, which got us tolerably warm." Smith spent a night nine miles from Prairie Bluff with the Honorable A. B. Cooper in whom he was very much interested. "He related many events of his public life. He is 86 yrs of age but tolerably active yet for so old a man. From him I got a MS Report with illustrative sections, made by Prof. Koch of his explorations of Monroe, Clarke, and Washington Counties. This I wished to have copied and bring it with me for that purpose and to be returned to him when finished." A lawyer, Cooper served many years in the Alabama legislature, helped reorganize the state after the Civil War, and took part in writing the first post-war constitution.

"Prof. Koch" was Albert Koch, a self-taught but knowledgeable paleontologist who explored Alabama twice in the 1840s to collect for his traveling exhibits. He focused his search on the prehistoric whale commonly known as the zeuglodon that had long interested Smith. In the spring of 1845, Koch had achieved his goal at a dig in north Washington County, Alabama, where he unearthed the skeleton he wanted.

At the close of this trip, the geological wagon traveled through a belt of longleaf pine woods to Butler Springs, a once popular summering place. Not only was it located wrong on the map, but the buildings were in ruins; even so, Smith noted, some families continued to spend their summers there.

Smith extolled the wild, rugged terrain in the Pikeville area—the steep hills, deep ravines, and hundred-foot high sandstone cliffs. Daniel Langdon is in the foreground. July 17, 1887.

1887–89

BUTLER, FAYETTE, MADISON, MARION, SHELBY, TUSCALOOSA, AND WILCOX COUNTIES

Chisholm's Orchestra toured in north Alabama—Florence, Sheffield, and Decatur—in July 1887, but by early September Smith was prepared for his customary geology outing and headed toward the Fayette Courthouse by way of Pickens County. Of particular interest this trip was clay that could have some commercial use. Near Sanders Ferry, Smith examined a clay smooth enough to use for white wash, then in a gully near the fork of the Carrollton and Columbus roads he noted some white and red mottled clays called "Castile soap" by the locals. Later, gullies around Pikeville showed a variety of good quality clays. "Across Beaver Creek the purple clays crop out on the hills and when worked up have yielded a deep red colored mud."

Smith photographed Daniel Langdon (right) and friends playing tennis on the University of Alabama campus in 1888.

He marveled at the abundance of gravel "at altitudes differing by 1000 feet and more" and puzzled over which formation (Quaternary or Tuscaloosa) it belonged to.

Smith's close associate for the last eight years, Daniel Langdon, had signed a contract with the university in June of 1889 as assistant professor of chemistry. Since the faculty was paid in twelve monthly installments, he drew two months' salary [$100 total] without teaching any classes because of summer vacation. Then he resigned his position. The Board of Trustees demanded he return his salary. Langdon refused. After the board's judiciary committee decided the dispute in Langdon's favor, he left Tuscaloosa, eventually becoming consulting geologist for the Baltimore and Ohio Railroad. He and Smith kept in touch through the years, and Langdon sometimes joined the later geological tours.

In midsummer of 1889, Landon, Smith's seven-year old son, fell from the top of a residence under construction at the university, striking his head on a heavy beam, the local paper reported. "His side was painfully bruised but no bones were broken and yesterday morning he was getting on very well." Three weeks later Smith, accompanied by Eugene Jr., left in the wagon for Coaling. At Keene's Mill, they were caught in a heavy rain and had to overnight in an old house. They moved on to Coaling where they set up camp, and Dr. Smith sent back to Tuscaloosa "for some easier shoes." Meanwhile he and Eugene Jr. examined the coal outcrops in the area.

On August 11, Smith stopped by the marble quarry on the Cahaba River bank that he last saw twelve years before. He was surprised at what had been accomplished and praised the marble's appearance and quality. In Montevallo on August 18, they made camp at the spring southeast of town despite the crowds that had gathered for a baptizing.

On August 21, they retraced a previous day's route because Smith was not satisfied with the conclusions he had drawn regarding the coal-bearing strata. The sandstones and shales near Calera confused him even more. His observations "all make it extremely difficult for me to be sure of any of our attempts at classification." Near Buxahatchee Creek out from Calera, he noted the occurrence of "a bed of limestone of variegated colors, white, red, and greenish,

Though Smith regretted erosion, he studied the layers revealed. He photographed this area near the Black Warrior River bridge in Northport, Tuscaloosa County, in 1888.

overlaid by a dark red slaty sandstone . . . This marble takes a beautiful polish and the colors are exceedingly pleasing. It ought to be better known."

Southwest of the Gap on August 24, Smith mentioned the settlement called Lick Skillet and drew a diagram of what he thought to be the structure of the Calera basin and the surrounding area. However later he wrote on this journal entry: "All wrong—see notes of Aug. 27th." On that date he wrote: "I tackled the Calera lime basin and its surroundings again and think now that I have gotten about the straight of it."

Smith's September 9 journal account summarized how the McConnico brothers, Allen and William, had guided him during the past week in his search for several kinds of fossil shells. He collected no satisfactory Turritellas but the brothers promised to send some to him in Tuscaloosa. The disintegrating Nautilus fossils created a sandy black soil that attracted cedars and grew corn well, but made cotton crops susceptible to rust. Another day he inspected W. S. Purifoy's crops on which he used greensand that he mixed ten parts to one of manure and grew "cotton stalks 6-8 feet high, with dark green foliage and loaded with bolls."

While traveling through Wilcox and Butler counties, Smith heard of a meteorite that fell in May 1883 near Atkins. Late one night several fishermen on Cedar Creek were startled by the rushing roar like a speeding train and an explosion of lights in the heavens. Next day searchers located the meteorite in a field where, over six years later, Smith examined it. He found "the rock which fell from the skies" puzzling; it resembled slag yet was lightweight for its size—eighteen inches at the greatest dimension and eight inches at the least with a weight of about forty pounds. "To me it does not much resemble any meteorite that I have ever seen. The surface is very uneven with botynoid protuberances, and where a piece is broken off it shows a great number of air cavities or bubbles. The surfaces are glazed as if this had been melted . . . The story of the fall comes very straight and there seems to be no doubt about the meteoric origin of the fragment, though as I say it does not look like the meteorites I have seen." After taking samples to be tested by Dr. Clarence King, Smith arranged to have the rock removed to a nearby store until he sent for it. Sometime later Dr. King, director of the United States Geological Survey, concluded the stone was slag and had no meteoric connection.

1890

BARBOUR, JEFFERSON, MONTGOMERY, AND RUSSELL COUNTIES

During August 1890, Smith made two short forays, the first to Bessemer where he explored the valleys surrounding Mr. B. Grace Jr.'s cabin in the hills beyond Valley Creek. The second was to visit Giles Edwards, a Welshman in charge of the Woodstock Iron Company, by now the second-largest producer of charcoal pig iron in the world. Edwards guided Smith to a railroad cut that revealed the Cambrian shaly limestone, so much "crumpled and distorted" that Smith took several pictures of it. Edwards said the limestone "makes first class lime, slaking somewhat slowly but perfectly to a creamy powder."

The main trip of the season came in October when Smith, accompanied by Judge Johnson and Dr. J. W. Spencer, Georgia state geologist, left Columbus, Georgia, at two p.m. aboard the snag boat *Chattahoochee* on the way to Ft. Mitchell. Among the layers exposed in the bluffs of the Chattahoochee River, Smith noted sand with grains "as large as buckshot." In Ft. Mitchell, he and his companions outfitted two skiffs while waiting for Langdon to join them. Then they took to the river, making camp the first night on top of a bluff on the Georgia side where rain held them over until two p.m. the next day. They managed to get downriver to the G. Y. Banks Landing, where in continuing rain they took refuge in a warehouse the afternoon and night of the twenty-second. On the way to Chimney Bluff, Smith saw "a great thickness of cross bedded sands" that contained "numerous fragments of stems and leaves lignitized, and in many places large logs lignitized, some part filled with the borings of toredo . . ."

Between Bradleys and Roanoke landings, they explored a fine shell bed that Langdon had visited before. Smith wrote in his journal that the "shells are weathered out in beautiful shape, especially sharks teeth and bones of reptiles, and teeth of reptiles, plates of the turtles." They collected many specimens from there and from a lower ledge where they found "finely preserved oysters with peculiar hinge. We got a great many of these with both valves in contact."

At noon on the twenty-third they came upon Captain Stratford's barge tied up near the mouth of Cowikee Creek. The hospitable captain invited them aboard where they put up until the next morning, towing their skiffs behind

Railroad cuts were of particular interest to Smith. This one in Eutaw Sands, in Montgomery County, cuts through a cemetery, 1890.

the barge as they moved downriver. The geologists stopped off in Eufaula to visit and stock up on supplies. Continuing their journey by skiff, at a "very high bluff of dark blue micaceous sand full of beautiful fossils" they collected a prize Nautilus shell "showing the iridescent colors." They camped on the riverbank near Ft. Gaines, meeting the *M. H. South* for mail delivery just after dark. Dr. Spencer left the party the night of October 28 due to the press of Georgia Geological Survey business.

At Alum Bluff, a long, sixty-foot high bluff near Bristol, they collected many shells and a couple of well-preserved bones of the zeuglodon. They waited at Bristol to board the *Ellis* shortly after dawn on October 30 bound for Chattahoochee where they took trains, Langdon for his home in Mobile, Judge Johnson for Gainesville, Florida, and Smith for Birmingham (after having to spend most of the night in Flomaton.)

1891

Autauga, Baldwin, Barbour, Clarke, Elmore, and Escambia Counties

Smith's father, Dr. Samuel Parrish Smith, died March 28, 1891, at the age of seventy-six. As a child he had come to Alabama while it was still a territory. He received his medical degree in 1836 from Jefferson College in Philadelphia and set up practice in Old Washington, Autauga County. He married in 1840 and moved to nearby Prattville in 1843. Eugene Allen Sr. was the eldest of his seven children, only two of whom survived to full adulthood. Eugene wrote Dr. Spencer of his father: "He was an uncommon vigorous man till within a few weeks of his death . . . He was one of the kindest hearted and most generous men I have ever known."

Smith, accompanied by his second son Landon (now ten years old) and Landon's dog Bob, left in the wagon for an extended tour in July, 1891. The rest of the family followed them in their buggy as far as Taylorville where they picnicked together and then parted. The wagon headed south but heavy rains caused such delays that by the thirty-first they had only reached Marion where Colonel Murphee of the military school invited them in. By August 2, they were in Autauga County slogging through heavy rains toward Wetumpka on the Coosa River, which they reached August 6. The high and muddy river concealed its banks from Smith's prying eyes. They camped there, planning the next day to visit the phosphatic marls that they had heard about.

From Wetumpka they followed the Old Plank Road (which Smith rated "as good a road as could be desired") by way of Good Hope and on to Tallassee, passing by "many good farms, the houses all . . . built with chimneys of great blocks of gneiss topped out with brick. Most of the houses had fine large well made barns, both houses and barns of sawed plank." He was awestruck, as always, by the approach to Tallassee. Here at the waterfall on the Tallapoosa River stood the second-largest cotton mill in Alabama. Smith regretted that the sun was not in the right position for picture making.

On August 9, 1891, they arrived about five p.m. in Columbus, Georgia, after driving the wagon through coarse sand characterized by large spots that had "a white center and red ring around just like an old sore." At the Chattahoochee River on August 10, they tried to force their way through the cane thickets to catch the snag boat, but were too late. Next day after breakfast, they tried again and managed to board Major Robinson's boat, which promptly hung up on a sand bar. After an hour of effort with still no hope of the boat getting loose, they returned to the wagon and drove to the Cretaceous ledge at Broken Arrow Bend for Smith to collect. On the twelfth beyond Enoch, which was struggling to become a town, they "got a peck of weevil eaten corn and some very good water." Landon took a swim in the Hatchechubbee as well as in Little Barbour Creek with Smith monitoring from the creek bank.

According to Smith, when they reached Eufaula at midday August 13, "it was as hot as it ever gets to be." General Clayton's son Jeff met them with their mail and an invitation to overnight at the Clayton home. Next day they went to the Clayton family's old place in the country for the boys to swim in the fishpond. They left town on the seventeenth by way of King's Bridge where Smith checked out a lime rock with shells that he had been told about. That night they camped not far from the large Blue Spring—clear, blue, and cold—which gushed from the limestone depths. This trip ended shortly afterwards for Smith to have a couple of days at home before leaving for the meetings of the Geological Society of America and the International Congress in Washington, D.C.

By mid-September, Smith, Landon, and Bob, with others who came and went as time allowed, returned to the field once again collecting in the southeast corner of Alabama. "I have been much surprised at the beauty of the country," Smith wrote September 22, ". . . for it greatly resembles what may be seen in many places in our Paleozoic region red or yellow clay hills, fine oak trees and fertile fields." Dr. J. N. Clements recommended to Smith that the university lands in Geneva and Coffee counties be put in charge of an agent to help prevent illegal timber cutting and to facilitate sales of land to individuals.

On September 30, while Landon caught four fish in about four minutes from Mr. Dixon's pond fed by a blue limestone spring, Smith was observing their host's successful use of lime and marl on his crops. Next day in Brewton many friends and "a large bundle of letters" awaited them. Socializing did not prevent Smith from answering fifteen of the letters after supper.

Wagon repairs and specimen testing on October 2 delayed their departure from Brewton until noon. They headed for Flomaton by way of Pollard, but heavy rains and nightfall forced them to take refuge with Mr. Martin two miles short of their destination.

South of Perdido in "an uninhabited country without roads," they lost their way and wasted time and miles before being set right. They crossed over the worst bridge Smith had ever seen—"half the planks were gone and one side of the bridge was a foot or two lower than the other." They reached Bay Minette, "a pleasant looking place," by October 4 and camped four miles south on the Daphne road in "a beautifully level piny woods," which had formerly been a turpentine orchard. Burned and dead trees surrounded them, and ponds and lime sinks appeared frequently along their route. Smith and Landon took a "delightful bath in the Bay from a bathing house on the pier leading out to the boat landing."

By November their way led north towards Jackson. As they passed the Spring Bank Church, Smith noticed walls in the graveyard made of the Grand Gulf rock that had been taken from a dig within a fourth of a mile of the church. In this area, he had trouble with the eastern Old Mobile road that was impassible in places, causing the geology rig to go out of the way several miles. He summarized the countryside as "one succession of crawfishy pine flats with grass, stumped pines, a few black jack shrubs and gall berries in spots."

At Mr. Patrick's, where they stayed overnight, six persons stayed in one room while Smith and Bob the dog slept in the adjoining kitchen shed. Reaching Jackson on November 13, Smith went to view the gullies at the edge of town and listed their layers in his journal.

1892

BIBB, FAYETTE, HALE, JEFFERSON, MARION, AND TUSCALOOSA COUNTIES

In January of the New Year, a fourth son was born to Jennie and Eugene. They named him Merrill Pratt, a name often used among the numerous Pratt family.

Meantime Smith was involved in preparing a collection to send to the mining exhibit at the Chicago World's Fair that the *Chicago Herald* extolled in its March 21, 1892, issue. The huge, sixty-foot wide gallery in the mines building was lined with 156 seven-foot high cabinets waiting to be filled with choice specimens from universities all over the country.

Dr. Smith and his children before Merrill Pratt Smith was born. Rear left is Eugene Jr.; front, left to right, are Landon, Julia, and baby Truman.

The following month, Smith was host to Professor Lester F. Ward, of the United States Geological Survey. On April 11, the two men took the train for Stewarts Station, southwest of Tuscaloosa even though the Warrior River was in flood and covered much of the track. At their destination, they hired a horse and buggy and drove to Havana to explore the Rocky Glen where Miss Tutwiler discovered the rare ferns, *Aspl. ebenoides* and *Trichomanes radicans*, then they jaunted down the Greensboro Road to investigate the "Caves" Gullies.

By the time they returned to the station, all trains to Tuscaloosa had been cancelled due to high water. Smith and Ward walked to Akron, about ten miles, where they ate supper and waited for the one a.m. train to Birmingham by way of Selma. After breakfasting in Birmingham, they spent the day "out on the mountain." With still no train to Tuscaloosa, the geologists took the short Mineral Road to Woodstock to stay overnight and next day hired a wagon to take them to Tuscaloosa. They progressed slowly, studying all the cuts in the railroad along the way. Of particular interest to Smith was a great arch that he had often seen from the train window. Five yards high with feet planted more than thirteen yards apart, this formation stood near the Bibbville station. "A crust of iron oxide, in places pretty fair ore, marks the outline of this arch very clearly," Smith discovered. He sketched it for his journal.

On April 14, Smith and Professor Ward took the geology wagon to Sanders Ferry where they "parleyed" with the ferryman who was apparently reluctant to take them across the flooded river. However, the geologists prevailed and after crossing ended up at the big gullies that so fascinated Smith. They spent most of the next day collecting leaf impressions and lignitized stems, "some very good things, better than at any previous time we have visited this place," Smith wrote.

Luck continued to smile on them. They made good time traveling over the Jackson Trail and the Mormon Road, finding them in far better condition than they expected. At the eleven-mile marker from the Fayette County Court House, Smith took test samples of what he judged to be excellent firebrick clay. They spent an entire morning digging out the "beautiful leaf impressions which perfectly fill these clays" followed by an afternoon of packing their collection and loading it into the wagon. They moved to the brink of the Sipsey Swamp where they camped, "afraid to try to cross it in the fast failing light."

At Glen Allen next day, Smith had difficulty finding a fossil plant site that he had discovered on his last visit because the intervening years had changed Glen Allen from only a railroad workers' camp to a growing settlement with a depot. Finally locating the site, Smith left Ward collecting and walked back more than three miles to fetch the wagon. By 5:30 p.m., he returned in time to help wrap up and pack in sacks "the immense collection" Ward had made. They loaded it onto the wagon and sent the outfit on the way back to Tuscaloosa. After having lunch and coffee, Smith and Ward caught the train to Birmingham where they arrived about eleven p.m. Back in Tuscaloosa by April 18, Ward left for Washington on the evening train "after having a most successful trip and making what he calls a grand collection," Smith wrote in his journal.

Mid-September found the geologists at old Sardis church near the headwaters of Big Sandy Creek digging a bounty of fossils at the mill. Another short trip at the end of September took Smith to Mr. Enslens's on the Montgomery road where he rode over acres and acres of exposed marl. Smith had already written a public service bulletin praising the virtues of marl and quoting a Smithsonian Institution geologist citing the Alabama farmer who covered "an old worn out, poor field" with three-fourths of a ton of marl, "no other fertilizer, with result of 1200 lbs. seed cotton. Before, it took average of five acres to make one bale." The Smithsonian geologist concluded, wrote Smith, that Alabamians have "one of the greatest mines of wealth of any people in America. In N.J. it takes 10–12 tons of the best marl to give good results. In Ala. ¾ ton did it in three separate trials."

Smith took time out to mail a fountain pen to twenty-year old Eugene Jr., who had gone to Prattville to work for the Pratt relatives: "I have used [the pen] for a while and found [it] to be a very good one. You must take care to put it in your pocket with the small end up or the ink will run out through the two small holes you see near one end. If the pen should let down the ink too fast so that it drops off and makes blots, you can get your Uncle Dan to tighten it up. . . . If you find that the pen will not do I will get you another . . . Your Uncle Will did not say how much they can give you for your work; that will depend in great measure upon how well you do . . . Do everything you have to do the very best you can, and never slight over, or half do anything. By your asking about the Christmas money I suppose you want some. I can send you a little if you need it and will let me know . . . Tell Aunt Julia that I think I shall be able to come over to Prattville some time about week after next to see about the monuments in the grave yard." The letter was typed, but vertically in the left margin Smith wrote by hand: "You ought to take more pains with your writing and read over what you have written so as to put in the words you leave out. You have left out a lot of them in this letter."

1893

BIBB, CHEROKEE, CLAY, CLEBURNE, COOSA, ELMORE, JACKSON, JEFFERSON, LAWRENCE, MACON, SHELBY, TALLADEGA, TALLAPOOSA, AND TUSCALOOSA COUNTIES

Apparently Eugene Jr. neglected to answer the letter. In February of 1893, Smith again wrote to him: "Your Uncle Will has told me that you are getting along very well, but I should like to hear directly from you about your work, what you have to do, how well Mr. Mims is pleased with you, and how much you are now getting a month. You had not been paid when I was last in Prattville and so I have never heard how much you got. Do you need any collars, if so I have a number and can send you three or four. They are no. fourteen and a half which I think is about your size. The [university] boys are all going up to Birmingham tomorrow to play football with the team from Auburn. . . . The public school has a holiday also as it is Washington's birthday."

Two events important to Smith occurred in the spring of 1893. Athletics, one of his great interests, had become more prominent in the life of the university, and on May 5 the school celebrated the second annual field day with Smith serving as a judge. At commencement, always the high point of the year for him, the university awarded to his friend and associate, Charles Mohr, a rarely bestowed honorary Ph.D.

The summer tour began July 18. Smith, son Landon (and probably Bob), left Tuscaloosa in the wagon by way of the Plank Road, camping at the summer house that night at what is now Fleetwood. Robert Jemison had built the Plank Road of planks twelve inches wide and two inches thick, cut in his sawmill and laid by his slaves. Next morning Smith spent several hours at Brookwood concentrating on the washer and the hoisting plant. "The washer is one of the Stein patent and seems to do the work very well," he noted. "Both the University and Peterson seams are mined, one on the one side, the other on the opposite side of the washer."

Smith headed for Vance, intending to go to Woodstock, but took a wrong turn and ended up almost to Bibbville before backtracking to set up camp. He had no idea where he was, and the people in the only nearby house had just moved in and did not know where they were either. Next morning Smith found out from a passerby that he was on the right road. The wagon then proceeded to Helena where Smith's attention was caught by New Town, a settlement of "nice cottages" no one had ever lived in because litigation halted work at the mines.

They spent the night at Mr. Squire's where Smith's mail had collected. There he learned that Daniel Langdon would be joining him in a few days. For some reason this was unsettling news: "All this deranges plans so much that I hardly know what is best," Smith wrote in his journal. At Montevallo on the twenty-first, they camped at the usual spring even though Mr. Vandergrift offered them his hospitality. On the Ashville road next day, at the first lime kilns, they crossed the fields to the old Tichenor place to see the big blue spring, remindful of the Florida springs Smith had seen—except this one was somewhat murky.

On July 26, they traveled to the Shelby Iron Works over rocky, hilly country where most of the longleaf pines had been cut for charcoal. Smith saw the steam shovels at work on the ore banks, the washer, the furnaces, and their prized patented roasting furnace. The geology group camped in a nearby grove and after supper returned to the works for Landon to see the 7:30 p.m. iron run.

By way of Shelby (the town), they crossed the Coosa River next day after a delay—"ferryman gone afishing down the river," Smith wrote. A quarry they visited drew its water from a big spring, pumping all it needed for the washers through a twelve-inch main. Smith noted regretfully, "The charcoal burners are beginning to denude these hills and in a few years they will be if possible more desolate than they are now."

While camped at Talladega Springs, Smith wrote letters and endured an afternoon of rain. On July 29, he and Daniel Langdon rode a wide swing from Marble Valley to the Narrows to the Wood's ferry road nearly to the river, then returned to Talladega Springs with Smith concluding, "I don't see how we are ever to separate the Kahatchee hills slates etc. from the semi metamorphic Talladega series. If they are not the same they 'deceive their looks.'" Still camped at Talladega Springs on the thirtieth, Smith was deterred from making some needed side trips because "the sun was so hot and the air so close."

After breakfast on the last day of July, Smith and Langdon rode to the top of Sulphur Mountain and back, then hitched up the geology rig and drove to Hanover by way of Weogufka. "That night we had two of the most terrific rain storms I have ever been camped in," Smith wrote. "By inadvertence one of the curtains of the wagon was left hanging inside the wagon bed, and before we knew it water enough had dripped in to wet thoroughly one end of our roll of bedding, i.e., a strip the whole length of the blankets, etc. and one end of the pillows. This was disagreeable but we came through the night all right, but for severe headache on my part which compelled me to sit up in the wagon most of the night."

They went to see James H. Neighbors next day hoping he could show them around the area. Smith found out he was "afflicted with dropsy of the heart and . . . will hardly live through the summer." However, at "the pit from which he [had] dug for me the altered corundum crystals . . . I got my saddle bags full of very good specimens." On the way back to camp, as they passed the church being built by the local citizens, Mr. McAdory invited them to the outdoor dinner, which they accepted.

On August 3, Archie Bentley, in charge of road work, not only assisted Smith in finding a hill where he collected tantalite, but also told him of a field where he would find tin ore crystals. The geologists ended this day at Hissop in Coosa County where Mr. Darsey welcomed them to stay at his house.

Smith devoted most of the next day to beryl crystals, "some of them of good color and transparency, fit to cut . . . I went in the afternoon four miles down the Swamp Creek Road . . . where a number of very pretty beryl crystals have been picked up. Here as at Hissop large garnets well crystallized . . . are also found embedded in the quartz. At both places there seems to be good promise of feldspar in large enough masses to be of economic value."

Later, from the Coosa County beryls and Cleburne County gold, Smith had Tiffany of New York create a ring of bright twenty-two karat gold set with three golden beryl stones. A Birmingham newspaper praised the ring's beauty and stated: "It shows not only is Ala. rich beyond comparison in iron and coal, but she is rich in gold and precious stones . . ."

Now, however, in 1893 despite pouring rains, the geologists went to see the Volcano out in the hills northwest of Hissop. Here there was a large deposit of forge slag where no people lived and where no forge had ever been. Some five years earlier, the locals told Smith, this slag erupted out of a hole, killing a pine tree. He knew without doubt that the slag was manmade, but how it came to be there and how the pine tree died he could not say.

Returning to camp at noon, Smith changed clothes and ate dinner before riding out to Mr. Thomas's to see the rock he claimed to have extracted silver from and the blue mud from which he had gotten lead. Smith took a sample of each, but he knew without testing that neither contained silver or lead. Yet Thomas insisted that James Mitchel, a silversmith, had made a silver collar button from similar ore taken from Mr. Goggans's land close by.

Smith learned further that Goggans had sent Mitchel's button and a sample of the ore to a chemist in New York for analysis. The chemist's report that the button was made from coin silver and the rock contained no silver enraged Goggans and convinced him that "everybody else was crazy and he was the only sane man."

Thomas caught the fever from Goggans. Even though a number of trained men had told him the rock he showed Smith did not contain silver and even though he witnessed Mitchel, the silversmith, adding a silver coin to an assay, Thomas still believed Mitchel. He even went into partnership with him, building a smelter and working with the ore.

Smith had met Goggans before. The previous spring, he brought Smith "pebbles and fragments" he had found on his land, convinced they were diamonds despite the verdict of several geologists that they were quartz. Smith agreed with that verdict, but Goggans remained adamant that the rocks were diamonds and that one of them was "the missing eye of an idol of which he [had] read some account," Smith wrote. Goggans declared if Jesus Christ himself told him that Thomas's rock contained no silver he still would not believe it.

"It seems incredible that any sane man would be so fixed in a belief," Smith wrote in his journal, "but these cases are only two more instances of what I hold to be a distinct form of monomania, and one as worthy the study of experts as any form of insanity in the hospital. A man once bitten by this mania has never been known to recover his reason."

Smith had now been making his rounds over Alabama for twenty years. He had seen many changes as, on August 7, when he and Daniel Langdon "rode out past Old Socopatoy to Bradford and back . . . Socopatoy is practically obliterated, a church and a house of the old lot being all that is left. Bradford is dropping to pieces."

Camped at the Hillabee Creek bridge on August 10, 1893, Smith was shown the Ulrich place by Colonel Bluford Dean, whom he had met during a previous trip to Alexander City. Dr. Ulrich, who built the elaborate house before the Civil War, had a vault blasted out of the slates for storing the wine that he made from his extensive vineyards. Well-known for his hospitality, he later went into mining quartz and processing the gold, leaving a legacy of numerous pits and the site of his mill on the creek bank. Smith made a photograph of the wine cellar opening, but it did not turn out. After the tour they returned to the Hillabee Creek bridge for a picnic with Colonel Dean's wife and the rest of the party.

The geologists spent that afternoon with J. B. Page who was in charge of the Hog Mountain gold mine, but they could not visit the mine because of rain. They decided to stay the night with Page and hike out to the mountain with him next morning. Smith described the area as "a 400 foot ridge of quartzite chiefly . . . but along its foothills are the same smooth talcoid slates that characterize this gold belt . . . I think the ore here looks more promising than at any other place I've seen."

They left Hog Mountain by way of Goldville where pits from the old days lay everywhere about them. Farther along, the true mica-schists appeared, "very highly garnetiferous, and these garnets of the size of buckshot cover the ground by thousands all along the way traveled . . . by New Site, and Fosheeton" and back to the Hillabee bridge.

That night after dark they moved camp into Alexander City and spent most of August 12 "purchasing supplies and attending to mail and other matters." Probably one of the "other matters" was finding a dentist for Landon who had for several days suffered from toothache.

Recent heavy rains interfered with the expedition's progress in more ways than one. On August 13, they faced a usually small stream west of Flat Rock that had been transformed into "a raging torrent eight or ten feet deep." They backtracked and sought another route. Next day swollen White Oak and Crooked creeks blocked their intended visit to a mining camp near Blake's Ferry.

In the vicinity of Riddles Mill on August 16, Smith happened upon the summer home of an old friend, H. M. Burt, who invited the group to dine. Afterwards Burt's sons took Smith to an old forge site to make a photograph, but he had to leave pictureless as he forgot to bring the camera lens. Back at the mill, Jack Riddle showed him his family's gold-bearing quartz mine about a half mile distant.

The travelers arrived in Talladega on August 17 to have repairs made to the wagon and to buy necessities. The next day they visited mica pits and drove out the Ashland road to the Franklin gold mines, but they found the mill not working due to litigation. They paused for lunch at the Enitachopco post office then went with E. O. Hobbs to the graphite mine that was equipped with "a five stamp mill and a grinder for crushing the schist. The graphite is then separated by a variety of appliances, taking advantage of the difference in specific gravity of graphite and the silicates. The separation does not seem to be perfect for the graphite even when prepared for market contains still some considerable amount of ash. Work is suspended here also, for what reason I don't know." On August 20, they passed the road to a tin smelting works, but since it had been shut down for years, Smith forgot about it until too late.

Two days later they camped at the Shattucks' "beautiful spring" in Shinbone Valley where the Dug Down road met the Oxford Road. The Shattucks' son regaled them with stories of a cave in the nearby Talladega mountains. A brook ran through it, and white sand covered the floor, the rock walls, and the roof. The boy could not tell Smith how the brook left the cave unless it went underground into another stream flowing below the mountain.

On August 23 near the Needmore post office, Smith took samples of "a curious greenish rock mottled with red spots as if of garnets, but these red splotches show some cleavage resembling feldspar." The next day they stayed over in Heflin because of an ailing mule. Smith used the time to visit a mica mine and to take Dudley, his driver and man of all work, to catch the train back to Tuscaloosa for a few days. In Dudley's place, Smith hired Ed Carter, a local youth.

On the afternoon of the twenty-fifth, the geologists toured the Echols mine situated upon a mica schist ridge with a shaft about seventy feet deep into garnet-filled sandy red loam that had a layer of quartz every two to five feet. When Echols panned the loam, it showed what Smith judged to be a fair amount of gold for dirt taken at random. The geologists then drove to Arbacoochee to look over several gold pits. In the Mud Creek swamp, they saw "where the whole face of the earth had been turned over in the 'forties' by placer miners."

Ed Carter proved to be a poor substitute for Dudley. "On return [to camp] I found our new cook with a fire ready but nothing else, as he had never made any biscuit or prepared a meal in his life. So Landon and I set to work and helped him out so that we got a very passable supper."

Near Arbacoochee the morning of August 27, they set up camp at Mr. Houston's, then, with Mr. Phillips as guide, they went over Turkey Heaven Mountain to two mines, the Moss Back and the Lucky Joe. Smith judged the Moss Back to be in "a good state of preservation" while the Lucky Joe he considered to be "well appointed" with a "40 hp. engine and a ten stamp Fraser and Chalmers mill of newest pattern and the plates and fixtures . . . of the latest patterns."

They left Arbacoochee by way of the Tallapoosa River high bridge at Bell's Mills, "a very pretty place where all the houses are neat and well kept and most of them painted." On the night of the twenty-eighth, they made camp at Mr. White's and next morning went with him to his gold prospecting site. While they were gone disaster struck Ed Carter. He had the wagon ready to leave when the black mule had a kicking fit and got the double reins tangled in such a way Ed could not control her. "She forced the wagon into an apple tree," Smith wrote later, "which tore off the top of the wagon, breaking all four standards on one side and one on the other besides tearing the cover pretty badly." They spent the remainder of the day until ten o'clock next morning in Heflin getting the damage repaired. On August 30, Smith noted, "Dudley returned on the 9 a.m. train," and the camp settled back into its customary routine.

At Phillips's mine the geologists saw a vein of mica fifteen to eighteen feet thick. "A long open cut and a tunnel have been cut into this vein," Smith noted, "and a lot of mica has been taken out. The house in which the mica was stored was burned down some time ago and a large amount of good mica ruined." Phillips also showed them a prehistoric pit overgrown with large trees that he opined was dug by Indians searching for kaolin rather than mica. At Pinetucky, Smith once again went down the mineshaft that he last explored nineteen years earlier with former Governor Smith as his guide. "They have done considerable mining since that time," he wrote.

On September 1, after safely fording the swift and deep Little Tallapoosa seven miles below Matilda at the Crossroads church, they found "outcrops of soapstone, quartzite, gneiss, and a curious green rock with pretty well crystallized green mineral." Smith recognized the latter as the same rock he found in this area years ago that had been identified as hornblende. Because it was so different from the usual hornblende rocks and because the sand surrounding it was "a peculiar light olive green," Smith believed it to be chrysolite but reserved judgment until he could test it.

In Dadeville on the third, they called for their mail and set up at an excellent campsite in an enclosed grove. Smith, accompanied by young Landon, spent some time driving about the countryside with a rented horse and buggy, analyzing the ridges and layers he observed and making sketches and notes. "In crossing Hatchet [Creek] between Brownsville and the station we got into rather deep water which came up into the buggy and wet thoroughly one valise and its contents as well as the rock specimens. This afforded no little amusement to Landon who has been giggling over it ever since. Returned to Dadeville by 8 p.m.," Smith recorded in his journal.

Leaving for Farrar's Mill by way of Walnut Hill and Barnesville on September 5, Smith observed at the Blue Creek crossing the same peculiar green rock he collected at the Crossroads Church. From the mill, they headed for Phillips's house but found the family away and not expected back that night. They set up camp "in the grove of an old church, our chance for water being very bad for we had an old well with bushes growing in it, and a spring which does not flow . . . Mr. Phillips came in after we had left his place and sent word to have us come down to breakfast with him which we are to do."

They spent next day with Mr. Farrar on Blue Creek. He dug ore from a hill above the mill, out of "a great backbone of quartzite," and processed it in his primitive works. Smith wrote: "[Farrar] . . . is perfectly crazy about his ore and possessions generally and says $20,000 is the least that will buy him out. [He] . . . wanted to dig up the whole mountain for me to bring off and behaved like a mad man . . . It was two hours before I could get away and then only after taking on a load of about fifty pounds of ore of every kind. Mr. F. has cut some fifteen feet or more in depth at one place, and 20 to 30 different holes where he has from time to time dug out a piece of quartz and I must see every one of them and he assured [me] that each had the best ore in the state. His mill does not always turn out according to his expectations and he accuses his neighbors of having 'greased' his mine so that the 'quick' does not take hold. Then again he thinks that . . . he had pounded up a rat in his mill and thus 'greased' things. One of the miners . . . told him that rats would eat the amalgam and he told me that they had eaten up a lot of it for him. Mr. Farrar . . . probably has a good piece of property . . . The great trouble here is want of water for in summer it is impossible to get water enough to keep the mill going half the time."

After extricating themselves from Farrar, they drove "over the barrenest quartz ridges I have ever seen," Smith wrote. He took photographs "but a picture conveys no idea of the utter desolation which the hills themselves give one. It is very depressing to drive through such scenery and especially was it so this afternoon on account of threatening rain. . . . Parson Guthrie, the father of twenty children when I was here twenty years ago, has now twenty-six and has moved down to old Ufola from near Susanna where he then lived."

On September 7, they traveled by way of Equality, Central, and Sand Tuck to Wetumpka, a distance of about thirty miles. The rain and the cold combined with a stiff northeast wind convinced them to take rooms in the Hotel Kyle rather than set up camp. They arrived in Prattville on the eighth where they planned to rest for a few days then separate: Dudley taking the wagon back to Tuscaloosa, and Smith leaving for the Chicago World's Fair with "the children and young people."

Smith's second decade as state geologist continued the progress of the first. He took great pride in presenting the university and the state to Alabamians and the world through the exhibits he prepared for expositions, and in guiding visiting geologists around, ensuring that they found in abundance what they came for. Enhancing the status of the science of geology remained high in his priorities. One means to this end was the Alabama Industrial and Scientific Society, which had been founded in 1890 by Smith and the chief mining engineer of the Tennessee Coal and Iron Company, Erskine Ramsey. Further validation of Smith's efforts came from the state—the annual appropriation for the survey was increased to $7500 in 1891.

Since he became state geologist, Smith had corrected mistakes on the seriously flawed maps he had to use. His twenty-year effort was published in *Berney's Handbook of Alabama* in 1892, a map of Alabama so exactly accurate it endured until 1926 as the standard work of its kind.

At least thirty-four publications came out from the survey pertaining to different aspects of geological and agricultural Alabama with Smith writing most of them, co-authoring others; overseeing others (i.e., McCalley's report on the plateau region coal measures and Squire's report on the Cahaba coalfield). Subjects included general agriculture, ores and minerals of industrial importance, the Tuscaloosa formation, phosphates and marls, gravels, and clays. Charles Butts, United States Geological Survey, rated Smith's paper on underthrust folds and faults (1893) as his "most original contribution to general geology."

3

The Third Decade, 1894–1903

1894

Bibb, Chambers, Cherokee, Chambers, Chilton, Cleburne, Coosa, Elmore, Lee, Randolph, Tallapoosa, and Tuscaloosa Counties

Eugene Jr. received a scholarship ($69) to a commercial college and left home in January for Lexington, Kentucky. In a February 14 letter to him, Eugene Sr. enclosed $5 "to tide him over."

In July Smith was again at McConnico's plantation and mill on Pine Barren and Prairie creeks where he found "three or four shells not obtained before from this horizon." He made two photographs of the McConnico house.

Shortly afterwards he and Landon left Tuscaloosa in the wagon at three o'clock, reaching the summer house (Fleetwood) at 6:30 where they set up camp. Next day they visited Tuscaloosa County mines in the Brookwood area, then went on to Milldale where one of the mules (likely John) seemed colicky. Before they reached Woodstock, the mule was "doubled up and apparently suffering greatly." They unhitched the mules and "began to doctor, trying everything we could think of but without giving much relief up to date ten p.m." Next day the mule was still alive but not well enough to work. They tied him to the back of the wagon and hired a horse to take his place. "At Woodstock we hitched him up and we came out to Greenpond early in the afternoon but he was so much under the weather that we pitched our tent and concluded to hold over." Three days later when he still had not recovered Smith decided to take Landon and go to Tuscaloosa for the election. They stayed two days, returning to Greenpond with Jennie and the other children who continued to Blount Springs where they customarily spent part of each summer. At this time the resort was owned by the Sloss brothers of Birmingham and was in its most elegant period attracting the cream of Alabama's society with its "Red Sulphur" water, its French chefs, the gambling and horse racing.

Smith and Landon, on their way to Jemison, saw near Clear Creek Crossing "a great abundance of large pebbles many as long as six inches and I measured one that was nine inches longest dimension. I don't see how such pebbles could have been transported by ordinary currents. They remind me more of a beach."

Upon their arrival in Jemison on August 9, J. D. Hand invited them to dinner. His house, Smith noted, was built "directly upon a chert ridge of Knox dolomite, the chert being penetrated the whole depth of a well 48 feet." They spent the afternoon at Hand's gold mine operation five miles west of town: "We pounded up some of the ore and panned it with result of showing a little gold . . . They have some analyses of the ore from here one of which shows over $2, another $5.60 and still another $15 or more a ton. None of the specimens examined by us this afternoon would show as much as the average of the three above."

Two days later, Smith and Landon "rode out across Yellow Leaf by Mum's Mill . . . in a gap of the mountain where old Mr. James Chand[l]er lives and whither he directed me to see some of the 'best iron ore in the country.' Among the fragments of quartzite which make the surface on the mountain I found . . . a few lumps of ferruginous matter, hardly to be called iron ore and even if it were good there could hardly have been in sight as much as one hundred pounds of it."

From Wetumpka the wagon proceeded to Prattville. While Landon stayed with their relatives, Smith met in Montgomery with Governor Jones and others arranging the printing of his Coastal Plain Report. "The day in Montgomery was the hottest I ever spent anywhere and nearly overcame me." In the afternoon of August 14, he met W. M. Brewer, Alabama Geological Survey, in Elmore, where they waited through "the hardest rain of the season" for the wagon to pick them up. It came near dark, and they made camp in a grove.

Next day they toured Old Coosada "looking over the . . . historic ground on river bluff back of Miss C. B. Reeves's, took a photo of the house though the illumination was bad. We also went on a little further to see the grave of Mr. Walker, one of the first settlers of Coosada." Smith took a photograph of the Governor Fitzpatrick house, but he doubted it would turn out. In Wetumpka they bought corn for the mules, and Smith met with several men he knew. Then they continued to Enslen Spring to set up camp where Smith had camped two years ago—still "a very inferior place but about the best to be had."

Near Robinson's Ferry on August 18, they ascended "a tremendous hill of quartzites and Talladega slates [that] make the Devils Backbone Mountain which crosses the river and extends down to west and south for several miles." At the ferry, Landon swam across the Tallapoosa River. They came into Dadeville about five p.m. on a Sunday, but even so the postmaster came down to the post office to give them their mail, the storekeeper opened his store to sell them feed for the mules, and Mrs. Johnson let them camp in the same grove they camped in the previous year. "At Mr. Holley's I saw a piece of Indian pottery nearly perfect, about 8–10 inches long and five or six wide and high," Smith wrote. The Holleys had found the artifact in the field near their house and used it as a water pan for their chickens. Smith was so taken with it that he sketched it in his journal.

On August 21, "Just before going down to the river flats, part of the breeching gave way on Mr. Brewer's horse and he began to kick and charge but was stopped before doing any harm except to break the cross bar of the shafts [on the buggy] which we tied up sufficiently to allow us to go on." That night they camped at a spring beside a schoolhouse and a church four miles from Alexander City. The next day they braved the rain to take the buggy into town for repairs, and Smith visited with Colonel Bluford Dean.

Afterwards, the rain renewed in earnest forcing them to make camp at a schoolhouse only four miles out of Alexander City. The rain came down "in such torrents as to put out the fire and compel us to take refuge in the schoolhouse where we also slept." On the morning of August 23, "in a hard rain we packed up and drove out here to the Hillabee Camp Ground where we found welcome shelter for wagon and stock under the great shed." Though the rain continued Smith and Landon took a short ride to Hackneyville "over the gneisses and granites all the way . . . Mr. Will Vernon has a good large school at this place and we have talked with him a good part of the day." The next morning Smith "took two pictures of the shed and church, one with the school children, the other without them."

On August 26 at Heflin, Smith noted, "Mad Indian P.O. is incorrectly placed on the map . . . [It] is a mile or two further up [Mad Indian Creek] and not on this road at all." The next day, they camped just south of Piedmont. "About 5:30 as usual it came on to rain and we were pretty thoroughly wet and had to put up our camp in the wet in front of Mr. ______ by whom we were very hospitably treated, Mr. Brewer going into the house for the night which promised to be a very disagreeable one."

At Tecumseh Station two days later, they drove out to the bauxite banks in the afternoon, touring the works and collecting specimens. The next morning, they packed two boxes of specimens and were ready to leave when Mr. Perry arrived with an urgent invitation to visit his works and take dinner with him. Smith was surprised at the amount of work that Perry had done since his first visit, sinking a sixty-foot shaft with several tunnels branching off it, cutting into the six-foot high lignite seam "whatever it is." Smith added in his journal: "This is a curious substance, a sort of bitumen, impregnated clay which at Washer Bank [which Smith had visited earlier] is light gray but which burns with luminous flame and ... very smoky ... with a peculiar smell. From it they have gotten a little oil by distillation. It seems strange to see a white clay burn, but it does so, leaving a large amount of ash. The stuff [at Washer Bank] is exceedingly light and porous and reminds me of diatomaceous earth. Through it are streaks and small strings indurated bituminous matter. The clay seems to be impregnated with combustible matter. At Mr. Perry's the substance is black or dark brown, more like lignite, and in it are many fragments of carbonized wood."

Afterwards Smith revisited the Washer Bank and took two photographs. They left Rock Run at 4:30 and reached Bluffton by 6:45, where they camped the night of August 29. Smith lamented the fate of the town: "Bluffton presents now a most melancholy spectacle of costly houses, utterly deserted. Buildings fit for a town standing out in the weeds with no reason for their existence."

September 1, 1894, near Oak Level they went with Joe Roberts "to see some pottery clay used by him and others in the manufacturing of jugs in the good old ante revenue days." Smith wrote of the extensive area of large pine trees between Mr. Roberts's and the Georgia line. He spent the afternoon "in writing letters, reading proof, and correcting the proof of the map. This latter I find the most difficult thing to do, as I have no facilities for making the corrections and no maps to refer to."

On September 4 at the Omaha store in Randolph County, "Mr. Ballard showed and gave me a large lump of limonite which when broken open proved to be hollow with the cavity lined with pyrite and the interspaces partly filled with granules of pure yellow sulphur, the first of the kind I have ever seen."

Continuing southward the next day, Smith wrote, "The flat rocks about Rock Mills are so massive as to show very little bedding." The geologists lunched in Roanoke, shipped specimens, and Smith wrote letters. He took time out to go with Mr. Hanley who "showed me around[,] introduced me to a number of gentlemen, and took me to his springs which are fine sights. They spring from the granite and are clear and pure." They camped at High Shoals at the house of the storekeeper and postmaster, Mr. Ussery. Smith viewed the falls on the creek and photographed them next morning, promising to send prints of the falls and the mill to Ussery.

Smith wanted to visit W. P. Spradling's soapstone quarry in Chambers County, but there "came on a heavy rain as usual just after we had finished our dinner which put it out of the question to go down into the weeds and bushes to the quarry. A great quantity of the soapstone had, however, been hauled out to the road for building purposes and we had ample opportunity of seeing what it was. It is of light greenish color . . ." He made note of several seams of kaolin in the area of Tuckersburg. As before when Smith had visited Oak Bowery, he praised the "many fine old residences . . . but most of them have been allowed to go to ruin." He listed the names of the old families: Ballard, McCurdy, Andrews, Ross, Kirk, Avery, Thomas. They camped at Kincheon Haralson's "in a fine grove."

In Cusseta the geologists picked up their mail and made camp on Dr. Lawrence Smith's grounds. They were treated "very hospitably . . . by him and Mrs. Smith, taking our meals in the house." Professor Smith spent the afternoon with Dr. Smith who visited one of his patients about eight miles below Cusseta near Halawaka Creek. Smith noticed how similar the region was to "some parts of the Canebrake country except in the colors of the soils which are not black [but] in some cases a very dark brown."

September 10 brought them within sight of the Chattahoochee River not far from West Point, Georgia. The old settlement of Berlin on the Alabama side of the river intrigued Smith. Its post office was named "Osanippa or Ossasnipper as some call it, and O'Snapper in the language of others." At Beulah that night, they camped in a church grove. "Before we had supper the people began to gather for the preaching, Tumlin of Georgia being the preacher, and we had a long string of boys and young men standing and squatting about looking on while Dudley cooked. After supper, we went into the church and heard a very good sermon."

Near Mott's Mill, the soil had a "crawfishy look" to Smith. On September 16, they visited General Pearson where they "got the full benefit of the scuppernong grape vine in his yard." In the afternoon, they packed up specimens and expressed them, then moved on to Griffin's Ferry, which they reached "about five, but having to make two trips to get our crowd over, it was nearly dark before we got out of the sloughs on east side of the island by the mill . . . The [ferry] is only 30 ft. long, without aprons, and we had a great to do to get the mules over."

Two days later Smith, at Cowpens, viewed the Rebecca Mountain "looming up in the west, plainly very much higher than any of the intervening country." By the twenty-first they reached Brownsville, a town interesting to Smith because it "was peopled with a Scotch colony, the descendants of whom now live about all this section. The names of some of the families are Brown, Carmichael, McDearmid, McCoy (called McKay), McMillan, McNeil, McLeod, McPhail, Evans, McConathy etc." Near Graham's Quarry, Daniel B. Brown, a surveyor, reminded Smith that Tuomey had mentioned Kennedy's mountain and several other local sites that contained enough magnetic iron to interfere with the surveyors' instruments.

Well on the way home by September 29, at Calera they saw "a large plant (now idle) for the saving of the distillation products from charcoal burning, and as many as 50 brick ovens have been put up. . . . At Montevallo I saw Mr. Vandegrifft who gave me a few gold quartz specimens. . . . And at Clear Creek I was informed that gold could be obtained from almost any of the branches and creeks. This must be confirmed."

They proceeded to Tuscaloosa by way of Greenpond and Vance arriving between three and four p.m. after a drive over "as bad a road as any we have seen this summer, badly worked, or not at all, and filled with rocks, the fragments broken up iron crusts, which could just as well have been thrown out of the road as left in it. In all the eastern counties where the stock law prevails, the roads are mostly excellent even when they go over mountains, but when we get into western Coosa the stock law fails and roads are generally neglected.

The same is true of Elmore, Chilton, Shelby, Bibb and Tuscaloosa counties. Not that the stock law promotes good roads, but stock law and good roads are twin evidences of a certain degree of civilization, which does not obtain in the piny woods districts."

1895

CALHOUN, CHEROKEE, CLAY, CLEBURNE, COOSA, AND TALLADEGA COUNTIES

Smith's father in law, the Honorable Landon Cabell Garland, L.L.D., chancellor emeritas and professor of physics, Vanderbilt University, died in February at the age of eighty-four. Jennie's mother had died some time earlier, and Dr. Garland was living with Jennie's sister in Nashville. The *Christian Advocate* wrote that he had spent nineteen years in Alabama, eight as professor of physics and astronomy at the University of Alabama, nine as president of the university, and two as president of the NE & SW Railroad. "He loved all animal life, was the avowed friend of every good dog and felt deep interest in birding." In addition, "he performed well on the flute."

In June, Julia Smith, seventeen, presented a piano recital and graduated with honors from the Tuscaloosa Female College. Her parents gave a lavish

The Smith garden on the University of Alabama campus was a wonderland of plants, many of them collected by Dr. Smith. In this photograph, Jennie Smith relaxes in the yard with sons Truman and Merrill.

afternoon lawn party for her at their home that "presented a scene of indescribable beauty and brightness," wrote the news reporter. Many amusements entertained the guests including "target shooting, tennis, [and] fortune telling under a gypsy tent . . . The charming debutante received and entertained her guests with easy grace; and those who were fortunate enough to be present will long remember the delightful event that ushered into Tuscaloosa society a young lady who is destined to be one of its brightest ornaments." Another newspaper reported that Julia "was charming in a picturesque costume of pale blue mull with a large black picture hat." Amelia Gayle Gorgas assisted Jennie with the serving. Among the guests was Mr. Yuille, one of Julia's most frequently mentioned friends.

Professor Smith spent part of July in the Trans-Pecos of Texas examining the sulphur deposits. He was back in the field in Alabama by August 13. Taking Landon with him he set out on the trail of the geology wagon and crew (including Dudley and McCalley), which he finally located near Jacksonville. They moved the camp to Piedmont where Smith and McCalley "got horses and rode out by Tampa again where we took shelter out of the rain, with Mr. William C. Cross who, among many others, has been the victim of a rogue calling himself Jacanthus Lawrence Fulton and pretending to be a member of [the United States Geological Survey] and of the Society of Mining Engineers. Some of his reports of mineral lands are rich reading. This fellow stole and swindled $300 away from Mr. Cross—married a girl . . . at Mark and has absconded. He is wanted at several other places for swindling."

The expedition was caught in heavy rain at Heflin just after noon on August 20 with no let up next day. "We therefore drove into Mr. Zaner's blacksmith shop (having some shoeing and other work to be done) and remained until after dinner" on the twenty-second. Then, despite a sick mule, they moved on to join Aldrich in Arbacoochee. Jim Hilton's "new find" at Arbacoochee was "a vein of quartz . . . some eight inches thick and for about a yard it has the edges thickly beset with gold, by far the most promising and the best specimen I have ever seen," Smith wrote. "One large piece in the store in Arbacoochie is estimated to contain between $100 and $200 of gold. A great quantity of nuggets have also been collected." The night of the twenty-third Smith noted: "Landon under the weather, ate nothing at supper and complains of headache and goes to bed early." Apparently he recovered quickly for next day the group returned to Heflin where they "camped in the bushes beyond Dr. Neal's. Mr. Aldrich, Gov. Smith, Mr. Scarborough and Mr. Winst. Smith were at the camp till train time when they went in towards Birmingham."

Smith's August 27 journal entry read in part: "Down below Buckeye we stopped at Mr. Hearn's and got him to pilot us up over Emory Gap to the Bald Rock and Pulpit Rock [on Mt. Cheaha] and a short distance across a ravine to eastward another similar cliff down which a small branch pours its waters in a beautiful Cascade." He made several photographs of the area. "We spent the day from 11:30 to 6 p.m. in making this tramp which we accomplished on foot carrying a camera on our backs. Returning we hitched up and came a mile or so down to Mr. Elder's to get feed for stock . . . Camped by the side of the RR one and a half miles west of Furnace."

On August 29, they went in to Talladega to visit Smith's Uncle James. The family was away "but the servants let us use the house for our dinner and helped us out with milk, preserves, etc." A great deal of mail had collected there for Smith; he answered most of it in the afternoon. They stocked up on provisions in town before going down to Dr. Taylor's for that night. Part of next day the geologists spent sightseeing with Dr. Taylor in a skiff, and Smith made photographs of the waterfalls, the cliffs and other points of interest. After taking dinner at Riddles Mill on Talladega Creek, they drove to Chandler's Spring for the night. Smith found several men he knew from Talladega, and had breakfast with one of them next morning.

Wide variation in the quality of water prevailed in this area. The water of Chandler's Spring had "a good deal of lime as well as of iron and probably some salt judging by taste . . . At Coleta the water is very strong of lime and curdles soap" while near Goodwater some days later "the water of the wells sunk in the blue slate . . . is always limy and unpalatable while the waters out of the mica schist hills is of the best." At Hissop on September 1, they lost two hours doctoring one of the mules. At Goodwater they set up camp at Mr. McLendon's whom Smith termed "obliging," but he added, "He lets himself be well paid however and is the only man who has ever charged for the privilege of camping by his spring."

On September 3, they explored the Indian fort, "a circle or ellipse of piled stones" measuring fifty yards across northeast and southwest and twenty-five yards northwest and southeast. "Near NE end [is] a very large pile of stones within the enclosure which is open on the SW end." Smith sketched it in his journal.

By September 13, they were in Columbus, Georgia, where Smith and Landon "took dinner at the Rankin and afterwards had a trip to No. Highlands by the electric cars. After supper drove two miles and camped." The next day, however, they had to retrace their way to Columbus because they forgot to buy candles.

Near the Hyde post office, they saw the field where the "old Indian council house was located . . . Mr. Meadows says the Indian chief known in the locality was called Tuscoonah and a little creek of same name runs into Little Uchee at the falls where Moffatt's Mill is now located." Smith took several photographs here.

After his return to Tuscaloosa, Smith was appointed to the Jury of Awards at the Cotton States and International Exposition to be held in Atlanta. The many prominent men and the few women from around the country who made up the jury gave two weeks of their time without remuneration. Alabama was also represented by Booker T. Washington and Charles Mohr. Washington's speech, delivered on opening day, was "printed in nearly every paper in the country [and] will be long remembered," reported the Atlanta paper. W. E. B. Du Bois termed Washington's speech the "Atlanta Compromise," the name one of the most important speeches in America has retained.

1896

Chambers, Chilton, Clay, Cleburne, Lee, Talladega, and Tallapoosa Counties

In the summer of 1896, eighteen-year old Julia Smith was caught up in a whirl of social events. On June 17, she attended a dance and reception honoring Vice President Adlai E. Stevenson (vice president, 1893–97) at the Tuscaloosa City Hall, which was elaborately decorated with red, white, and blue bunting, stars, shields, and eagles. Unfortunately, Vice President Stevenson was too fatigued to attend, but Julia and her escort, Harry Thornton, danced to the music of a Mexican band.

The following night, wearing "a tasteful dress of white organdie over pink silk" according to the local paper, she attended the Commencement Hop in the Old Mess Hall on campus where an Italian band played for the dancing. Two days later she was at the annual banquet of the Sigma Nu fraternity accompanied by Mr. Stevens.

Meanwhile, Landon, now fifteen, made plans once again to tour with the geology wagon. He and the expedition left a week before his father joined them in mid-July at Sylacauga. On the way to Hollins, they were caught in such a torrential rain that they had to bring the wagon and the buggy to a halt and wait out the worst of it. When the storm finally lessened, they made their way to Mr. Williams's place where the mules had shelter and the men took refuge in a nearby church.

Next day they inspected a "very complete" saw mill and studied a large-scale map of the area. One night William N. Brewer of the Georgia Geological Survey, a member of the expedition, went "to the house of Mr. Franklin on account of some reported trouble between Mr. F. and one of his renters." For two nights, the geologists camped at Franklin's. On the morning of July 23, they collected slate samples from the Jenkins place and "inspected again the beautiful Chalybeate springs." On the way to set up camp at the Idaho Mine where Joshua Franklin was superintendent, they checked in again at his home. Finding everything peaceful there, they continued to the Watts place where a Talladegan named Ivey was setting up a gold mill.

Later Smith wrote the *State Herald* that a paper Franklin presented before the Alabama Industrial and Scientific Society this year showed conclusively "that the low-grade ore in that section may be worked with profit." (January 18, 1897)

On Sunday, July 26, Mr. and Mrs. Franklin spent the day at camp with the geologists. "Mrs. F. helped the boys about cooking and the result was a good dinner," Smith wrote. Monday they went to Chandler's Spring for their mail and then visited the waterfall on Robeson Creek. After dinner they moved camp back to the Franklins' and Dr. Mohr joined them. "We botanized down the Creek, and Dr. M. found two things new to our Flora . . . After tea we talked botany with Mr. Franklin who is pretty well posted, till bed time."

On the morning of July 29, they trekked up Cheaha Mountain "going by a more SE side of the high peak, passing over it and down to the waterfall where we killed a large rattle snake (52 in. long and 6 in. dia 10 rattles) and took our dinner. Skinned the snake for Landon, and after dinner we went out to the Pulpit Rock then back by Bald Rock and down by the same route we ascended last year. The trip was very successful in a botanical sense, as well as agreeable in other respects. We took along two mules, one of which Mr. Mohr used, while the rest of the party took turns riding . . . Returned about 6 p.m. The mountain is made of the quartzite almost exclusively."

On July 30, they descended the mountain to the two thousand foot level where they explored the "Cave." From one opening, two caves branched off, sloping steeply and "diverging from each other . . . [reaching] a depth of 100 feet or so" from the main opening. They camped that night on the roadside. "About a mile from Delta," Smith wrote, "we noticed a fine peach orchard with abundance of fruit, a rare sight this year."

The geologists found the citizens of Delta involved in politicking, which was also true in Ashland when they arrived there just past four the next afternoon. They camped Saturday and Sunday on the schoolhouse grounds, where they had plenty of company. Crowds had "come to hear the joint political speaking and as a matter of course our camp came in for a goodly share of attention. I was only able to hobble about both days because of a bruised foot caused by a heavy scantling falling across the instep on Fri. just as we were driving into camp. Left town about 11 o'clock having waited for the blacksmith to make a new cooking rack, and came out the Graphite Road and camped at our old place, the blacksmith shop at Mr. Gravits."

On August 9 when Mr. Brewer left for Atlanta from the camp at Millerville, Smith and Landon accompanied him as far as some rock-bearing garnets which they examined at length. While in the vicinity of Ashland, Smith met W. N. Crawford, "an old miner and mineral man, with whom I went down the creek half a mile or so to inspect a shaft which he had sunk on the blue granite . . . Mr. C. thinks he is on the track of a vein of tin, for what reason did not appear." At Chamber's Spring, they camped "within the Springs enclosure where we had several young ladies and others to see the camp. After supper the boys and girls had a dance at the shed and platform."

Camped once again at Ashland on August 28, they went out to Mr. Joshua Franklin's for afternoon dinner even though Franklin was abed with a fever he had picked up in Ashland whence he had gone to appear as a witness in a trial.

In Tallapoosa County not far from Goldville, Smith "bought 50 ears of

Waterpower was a major interest of Dr. Smith. Here he photographed Giles Mill near Abner in Tallapoosa County. September 1, 1896.

corn and Landon a load of fodder out of a field one fourth mile down in the hollow between steep hills." Near Bluff Springs on September 3, Smith wrote, "Soon after leaving camp this morning and while the wagon was waiting for me to collect some samples . . . an ox wagon passed by. At this, one of the mules pretended to be frightened and jumped back and forth in such a way as to break out the eye bolt of the single tree and the strap of the breast yoke. To mend these cost us an hour's time."

On September 5 they were back at Franklin's to set up camp. Franklin was still bedridden though improving. Smith drove to Ironaton to meet Charles Van Hise, professor of metallurgy and geology at the University of Wisconsin, who came in on the 2:10 p.m. train. They happily "geologized" all the way back to camp. Smith took Van Hise to see Ivey's gold mine, then Franklin's, and on to Chandler's Springs where they camped. On September 9, Smith sent the wagon ahead to Chamber's Springs while he and the professor drove a buggy on a side trip to points of interest including "a bold bluff of the greenish silvery Talladega slates which have been beautifully crumpled." Smith collected "a number of fine museum specimens." By September 12, they reached Opelika where Professor Van Hise left for Montgomery.

The geology wagon then proceeded to Chambers County. Smith complained about the LaFayette railroad that they crossed "a dozen times at imminent risk of breaking our wheels because of no boards being at the rails." On to Daviston and Dudleyville and "beyond the soapstone belt" ending up once again at Holleys' where Smith bought for fifty cents the Indian soapstone bowl that had so caught his fancy two years ago, the bowl the Holleys' chickens were still using for a water pan.

Again in Ashland on September 16, Smith found many letters awaiting him. One from Mr. Hayes, United States Geological Survey, arranged with Smith "for a joint trip across the Anniston and Tallapoosa sheets. As usual," Smith wrote, "I get it a day too late and two days before I can get answer or word to him by wire." Hayes reset the date to September 18.

Meantime Smith took the wagon to "Flat Rock by Lee's Gap, thence up around by Bishop's on Shinbone Road then back to Mineral Spring for camp." On the morning of the eighteenth, he explored "the old copper mine shaft again, went into the tunnel as far as we could get, because of the water." At noon the wagon paused at Franklin's where Smith left it to go in a buggy with Mr. Brown to visit his gold mine; Smith judged "the saving of gold seems to be pretty fair under the [present] management." Here the wagon picked up Smith, and they returned to Ashland to await what he termed the "U.S. party," Hayes and those accompanying him.

When the visitors arrived at noon on the next day (Saturday), the expedition moved out to Liberty Church to set up camp. "A very good rain during the night laid the dust and Sunday a.m. the weather was delightfully cool," Smith wrote. He sent the wagon to camp at Shadix's Spring by way of the Shinbone Road while he took the U.S. party about the countryside in the buggy. (He and Hayes disagreed over the identity of the curly schist.)

While Smith was away from camp on September 22, Hayes went to the Heflin depot expecting to meet his wife. When she failed to arrive he continued to Anniston without seeing Smith again. Now just past mid-September, the weather turned unseasonably cool; because of scanty bedclothes, the expedition closed camp and went to a hotel. A few days later in Trickem Valley, Smith photographed Turkey Heaven Mountain from several viewpoints.

On the morning of September 28, they rolled out of Ashland leaving "everybody in town at the school house and all the stores closed" in celebration of school opening. They continued "by Hollins to the crossing of the Plank road but could find no corn for feeding and so came on a mile down the Stewartville road to Mr. Carlisle's before we could camp, bringing us up about good dark."

1897

BIBB, CLEBURNE, COOSA, CULLMAN, JEFFERSON, RANDOLPH, ST. CLAIR, TALLADEGA, AND TALLAPOOSA COUNTIES

In May Smith implemented a three-day tour of the mineral district for a group of senior university students. Loaded with the necessities for a joyful outing, including lunch baskets and cameras, the young men, accompanied by Smith and Daniel Langdon, took the mail train to Woodstock where they boarded a special car courtesy of the L & N Railroad. They visited the DeBardeleben Blue Creek mines where Superintendent Johns and his assistants took them down into the mines for several hours, "explaining everything

Group of Dr. Smith's senior geology students near Woodstock, Bibb County. May 13, 1897.

so clearly that the boys now know the combination exactly," according to a newspaper report. Afterwards the coal company treated them to an "elegant dinner, prepared for the occasion, and all partook with a hearty relish." The students toured the Bessemer Pipe works, furnaces 1 and 2 in Bessemer, the brown ore mines at Greeley and the red ore mines at Sloss, and the "fine quarry of limestone just across the valley at Gurley." At the latter place the students were "fortunate enough to witness several large blasts, fired just as they reached the quarry." Other points of interest were on their agenda, and at each place experts guided them and answered questions. Twenty students, all male, took the trip, returning to Tuscaloosa at midnight "tired and hungry but gay as larks," according to the *Tuscaloosa Times-Gazette.* The paper added, "The co-eds backed out at the last moment and wouldn't go."

The 1897 field trip left Tuscaloosa on July 28 with both Landon and Truman Smith aboard. They aimed for the Hardy Clements place at the headwaters of Big Sandy Creek going by way of Scottsville to Centreville past the sites of "Soap Hill of ancient stage times fame" and the old Bibb Court House on Copperas Creek, both mentioned by Smith in his earlier travels. By the thirtieth they were in the Weogufka area headed for Hollins but took time out for a swim in the Coosa River. Landon especially loved swimming; he grasped every opportunity to develop his skill.

At Camp Aldrich on Hog Mountain August 7, they panned for gold resulting with what Smith called "a good show" from every try. Two days later, camped at the familiar Hillabee Camp Ground, they all took a swim in "a good wash hole a little above the ford" probably in the Tallapoosa River. Next night they camped at Walker's Ferry stopping the wagon "on the only ten foot square of level space in the neighborhood." Landon set up his cot "right in the middle of the road under a tree. Our wagon had to take open sky." On August 16, they lay over at a plantation where Mr. Busby, "an old acquaintance" of Smith's, was in charge. They enjoyed his hospitality—the boys had the comfort of sleeping in the house, and the wagon's cook had the use of the kitchen stove.

At Daviston on August 18, set up in an old store, they were held hostage two days by torrential rain. Smith noted in his journal that "the roads have been fearful and almost impossible to pull through in places." They examined the graphic slates between Corn House and Wild Cat, and at Pinetucky Mine on August 21 set up "in a vacant house by the politeness of Mr. Fair who has

The survey camp, August 17, 1897, Tallapoosa County. Both Landon and Truman Smith accompanied their father on this field trip.

charge. No feed to be had anywhere about except 26 ears of corn which Mr. Fair spared us. We had fortunately some oats along and managed to make out with this."

In the area of Arbacoochee on August 23, Smith visited a "new gold opening . . . The ore is a laminated quartz, sugary, which is in a roll like some kind of jelly cake." Wood's Copper Mine that so impressed Smith in other years was now closed. He also visited Mr. Roberts who "was doing some sluicing, he having turned aside the creek and worked over the material in the bed of the creek. He has a lot of first rate nuggets, one of which [for] $4.50 I got for the University collection."

One morning Smith "went with a Mr. Tunnis out on the ridge of Talladega slate lying to the south of Gooseneck Valley. He has sunk a 90 foot shaft through the decayed slates with quartz lenses, upon a theory of his own. He made the location by means of a battery of his own invention. He is confident that he can disprove all geological theories heretofore advanced."

Headed south for Wedowee, they traveled along the banks of the Tallapoosa River for some miles. "Several of the rock bluffs that came down to the very edge of the water showed a beautiful folding and gnarling of the schistosity or stratification, whichever it is. I would give a good deal to have a five feet square slab of it at the University . . . We stopped at Wedowee Creek and took a wash before coming into town."

They spent the twenty-sixth in Wedowee "attending to correspondence, seeing the brick making for new court house and arranging for a trip out NW of town in the afternoon." By August 28, they were in the Roanoke area where James R. Gladney let Smith have "a large crystal of quartz weighing over 60 lbs., found on his land some years ago. I also got a good sample of titanic iron . . . from Mr. J. W. Smith . . . Mr. Smith is now digging for a copper vein . . . I don't think there are any signs of copper in what he had thus far brought up, but he is led on by the finding of some porous limonite which he takes to be

In September 1897, Smith photographed several views of Turkey Heaven Mountain from Trickem Valley. In the foreground of this September 1st photograph is a flourishing farm.

Landon and Truman Smith were with their father when the survey crew camped at Hog Mountain, September 23, 1897. Earlier, they had panned for gold, resulting in what Smith called "a good show."

copper gossan."

On September 1 from their camp in Arbacoochee, Smith again turned his camera on Turkey Heaven Mountain getting some shots that he rated "splendid." Next day he mounted a mule and set out to trace the Hillabee outcrop about Arbacoochee, but was suddenly seized by a severe attack of colic that compelled his return to camp. After several days of great pain, he rode the wagon to Heflin and took refuge with Dr. Neal. When he was able, Smith went to Anniston and boarded the train to meet the wagon in Ironaton.

At Waldo, Smith managed to keep several appointments, among them visits to the May Virginia mine and the old Story mine. "The impression left on my mind is that while at both these places, very rich ore occurs, yet the vein is thin and irregular . . . and hence there is very little prospect of profitable mining. However the enclosing slates are to some degree impregnated with gold and the old slate dumps have all been milled with some profit. Notwithstanding all this, there seems to have been very little if any profit derived from these mines by the owners though some of the country men who have prospected on their own account have made fairly good day wages with rocker, mortar, etc."

They stayed three days within Mr. Williams's familiar enclosure at Talladega waiting to hear from Truman Aldrich at Hog Mountain. "We had a good rest up from our usual camp fare and enjoyed the hospitality of our friends exceedingly." The following week the expedition crossed Choccolocco Creek and came to Riverside, choosing their campsite with care to thwart the roving bands of hogs. All the crew took "a good swim" in the Coosa River.

For the next several days, the expedition visited coalmines in the vicinity of Fairview and Ragland ending up at Lock 3, "a beautiful place" on the Coosa River. They took a swim "in the lock and above it," and feasted at supper and breakfast on fish caught by the lockkeeper. They "went up on the mountain" to look over Major Singleton's ore works. The major sluiced the brown ore, "embedded in red clay," downhill through "gates and screens" of his own invention, "a simple way of washing the ore but . . . apparently efficacious," Smith wrote. On the way up Beaver Valley to camp near Ashville, they were caught in a "heavy rain that lasted an hour or two, a very welcome thing since the heat and dust have been fearful."

The morning of the nineteenth, taking Truman with him, Smith left the camp at Leeds and went to Birmingham to meet with Aldrich and Colonel Montgomery and to buy provisions. On September 21, as they traveled to Hog Mountain, Smith made note of "the denudation of the country" by sawmills and the Shelby furnace. They stayed at Hog Mountain three days with Truman Aldrich,

Entrance to Aldrich's mine, Hog Mountain, Tallapoosa County. September 26, 1897.

William Brewer of the Alabama Geological Survey, and others, "taking notes on the veins exposed." In Goodwater on September 27, Smith got Landon and the wagon started for Tuscaloosa, then he and Truman boarded the train and arrived home that night. However, the 1897 field trip was not quite finished—as soon as they reached home both Landon and Truman took to their beds with chills and fever.

1898

AUTAUGA, CALHOUN, CLAY, COOSA, ELMORE, JEFFERSON, ST. CLAIR, TUSCALOOSA, AND WALKER COUNTIES

Smith, Truman, and their driver left home August 5 and set up camp about six p.m. beside the Methodist Church at Brookwood. Colonel F. M. Jackson, treasurer and general manager of the Standard Coal Company, was still "somewhat lame from the railroad accident near Bessemer the other day," so he delegated to someone else the job of showing Smith around the mines. In the first five years of operation, the Brookwood mines had produced over half a million tons of coal and about two hundred thousand tons of coke from the 202 beehive coke ovens "generally conceded to be the best built in the South," according to the *Tuscaloosa Times*. The company owned "over 31,000 acres of coal lands," the newspaper added, with ". . . enough coal . . . to mine . . . for over 200 years" at the rate of two thousand tons a day. "In addition . . . the company owns as fine a body of yellow pine timber, covering about 16,000 acres, as can be found in Alabama." The president of the mines, Henry L. Einstein, lived in New York City and had a summer home in New Jersey, the *Times* wrote. A young man, he "entertained royally" at his New York home. The paper described him as "prominent on Wall Street" and "a man of extended information and a clear cut, terse way in his conversation that is very attractive . . . He seldom comes south." During these years miners' wages ranged from forty cents to sixty-two and one-half cents per ton of coal dug.

On August 8 on the way to camp near Little Belle furnace at Bessemer, they "got into a mud hole with perpendicular sides and one or two feet deep and in pulling out broke the double tree. Patched it up and came on. The camp was in the midst of the Flatwoods shales and the water execrable. After supper Truman and I went up to the furnace . . . and saw them draw off cinder several times," Smith wrote in his journal.

He went into Birmingham to meet with Professor John F. Newson of Indiana University and Professor O. B. Perry of Vincennes University who were touring the iron districts. For an afternoon and evening, he pointed out to them important facts about the Birmingham area. When he rejoined the wagon, they camped at Colonel Montgomery's where Smith and Truman slept in the Montgomery house, and the driver slept in the wagon. "It poured down rain the greater part of the night but everything kept dry."

They took the wagon over Oak Mountain on "a well nigh impassable road" to Ragland "which we reached about dark and camped by the old grist mill across from depot on the bank of Trout Creek." On August 12, Smith rode west of Ragland with Mr. Brown to see a new seam of coal that he had just opened. Smith took a sample, and the next day while writing letters set fire to it on a sheet of tin. He found it slow to burn and its "crush structure . . . an unfavorable sign."

On August 14, Smith stopped off in Anniston for Truman to see the Pipe Works, and they dined with Judge Lapsley. In the afternoon, Smith wrote letters, had the mules shod, and bought provisions. Then "we drove down to the lake, our old camping place where we enjoyed a bath and a boat ride." Outside Arbacoochee on August 16, they visited a gold sluicing operation, and on the nineteenth in Chulafinnee, they saw "gold of a deep red color" being panned. Arriving in Ashland at five o'clock the twenty-first, they set up at their customary place near the schoolhouse. In the two years since their last visit, the blacksmith shop had burned and several houses had been built, Smith noted.

He hired a surrey for a circuitous trip out from Ashland. As they traveled over a new cut public road, the surrey's underpinnings caught on a hidden stump in the road breaking the doubletree. Smith patched the break with a piece of hickory from a nearby sawmill, and they made their way back to camp "all right."

The wagon lumbered into Goodwater about four p.m. on the twenty-fifth in a heavy rain that restricted their activities; nevertheless Smith observed the terrain well enough to confirm several of his previous observations. That night their driver and the mules slept in the stable while Smith and Truman stayed in the home of their host.

Moving on toward Deatsville, Smith expressed his concern about the weather in oxymoronic terms: "The east wind which has been blowing up for two days bears fruit tonight in a steady drizzle which set in as soon as we got into camp. The promise of an exceedingly bad night is remarkably good."

About ten miles out of Prattville early in September, Smith toured three potteries that used the Tuscaloosa clays of the area. Of a nearby ochre works, he wrote: "The material is regularly mined by entry and brought out by trains. Then it is spread out on shelves under cover and dried and then put through the various mills. Some of the mills are regular grist mills. The raw material is a yellowish clay with streaks of white. This establishment puts up the ochre in barrels for shipment and the work has been going on for a number of years." The ground up ochre, mixed with oil, was used in the manufacture of paint.

Of late Smith had come across some memorable gullies; several near Kingston exposed "pink and red sands and clays of Tuscaloosa age." Another, about a mile from Robinson Springs, was so large a bridge had been built across it.

In the northeast corner of Autauga County at Lock 2 on the Coosa River, they ate dinner with Mr. Wadsworth and then rode on his logging train to see his works on the river. He had a pond "about 75 feet above low water . . . into which the logs are elevated by an endless chain with spikes and from the pond to the cars 30 feet higher by the same appliance, run by a terobine wheel, and the water from the pond." Afterwards they hiked to Gold Branch to pan gravel from the banks, "getting a good show of gold in every case." Next day they went to Chestnut Creek to inspect the dam and sluice Wadsworth was building to store his logs.

On September 11, aiming for Oxmoor, Smith traveled along the crest of Shades Mountain. "On the second Conglomerate Hill, however, we failed to see a right hand trail, miscalled a road, and kept on down the trend of the hill till we came out at Brock's Gap. From here we turned up parallel with the railroad, but because of errors in the U.S. topographic map, which puts the dirt road on the wrong side of the RR, together with the dimness of the socalled public road which here is very little traveled but very much washed out, we blundered about for an hour or two and at one time had the prospect

This gulley at Kingston, Jefferson County, interested him because it exposed "pink and red clays of Tuscaloosa Age." September 1889.

of being overtaken by night in a region where no one lives and where roads can scarcely be said to exist. We finally got on the right track and camped . . . about two miles below Oxmoor."

On the way to Palos, they crossed the Locust Fork of the Black Warrior River. The road from the ford to Palos "was of the very worst description. If we had not actually come over it I should call it impassable." The man Smith wanted to see was away, but he went into the slope far enough to satisfy his curiosity. On the way to Horse Creek the next day, "we found one of the braces of our wagon tongue broken, and small wonder considering the strains to which it had been subjected."

At the Ivy Mine, they spent the morning seeing "the tipple washer and coke ovens, and [going] into the mines. The coal is some seven feet in thickness in three pretty well defined benches. . . . A characteristic of the Ivy Mine . . . is the vast number of flattened trunks of sigillaria-calamites and lepidodendrons which fresco the roof. One of these stems was exposed for at least thirty feet with a width of twelve to eighteen inches. I saw stems flattened with the leaf scars in double rows with unmarked spaces between . . . I don't know that I have ever seen anything just like it. Also in the roof in different parts of the mine there were numbers of 'pots' or pipes, sections of trunks ready to drop out and in places partly dropped out, a great source of danger to mines. I also saw a very long calamite fifteen or twenty feet and six inches in diameter, only partially flattened. This I shall try to get."

In addition, they visited the Price mine, which "had not so many tree trunks, but there was one 20-25 feet long and two and a half feet wide in the roof of one of the entries." On September 20 at Carbon Hill, Smith and Truman went down the Galloway mine, and later a mine at Pocahontas, both on the famous Jagger seam. In both mines, the "floor and roof are as smooth as possible and shots move forward great blocks of the entire seam, both benches, and from these the coal is easily broken up into large lumps."

1899

BIBB, JEFFERSON, ST. CLAIR, SHELBY, AND TUSCALOOSA COUNTIES

In May 1899, Eugene Allen Smith Jr., turned up as mailing clerk on the staff of the *Tuscaloosa Times*. His likeness, sketched by an artist, showed an unsmiling and clean-shaven young man. That summer Eugene Sr. took a fossil field trip to Wyoming for almost two months. His return to Tuscaloosa was timed to allow a week for taking care of his correspondence and preparing himself and Truman to take the geology wagon out September 6.

On the route to Wheelock Springs, Smith noticed "great changes along the road mainly in the abandonment of houses and the disappearances of others. The Randall Brown place has been completely destroyed by fire and after leaving Cottondale we pass no houses which are inhabited till we get to the Coaling RR when we find a settlement. I notice some changes about the Springs here

A large lump of Montevallo Coal. Unknown photographer.

but the water still comes forth in good volume and seems to be as good as ever."

From Blocton he intended going to Aldrich but discovered that route would require crossing the Cahaba River "at the Lily shoals [and] travel dim roads that are not worked." This he was not willing to risk despite losing time and mileage by the detour. Though late arriving at River Bend, they "got a fairly good camp under a hickory tree in the lane and at a convenient distance from the well which affords pretty good water."

At the coal mine west of Aldrich, Smith found that the miners sometimes loaded the cars with the dominant but inferior coal that underlay the desirable Montevallo seam. He visited several other workings but wrote later: "It was terrifically hot and sultry by the time we returned to Mr. Aldrich's about 3 p.m. and drove out . . . [to Dogwood]; the road was very heavy on account of the rain and the mules had a tough time of it. Dogwood Church Spring was a mudhole so we drove a few hundred yards further on and camped near a spring on the road to Cary's Mine. The water here also muddy."

At Cary's Mine September 15, Smith "took a sample of the hard coal from a large piece in the yard which showed the whole thickness in this part of the seam about 18 inches. In one of the rooms near bottom heading I took another sample showing the average of the Under Bench or soft coal part of the seam about 8 to 10 inches thick at this spot. . . . The soft coal is not put on the market as a rule . . . but is used under the boilers here."

The wagon left Dogwood on Sunday, the seventeenth, headed for Coal City by way of Maylene "up nearly to Siluria and out by Harpersville road . . . between the two parts of the Coosa [Coal Field]: i.e., between Big Oak and Double Mountain." Beyond the Big Narrows "we came the Vandiver Road down Bear Creek Valley which to my surprise was far more densely populated than the Valley of Peavine creek below the Narrows. Houses here were passed every quarter to half mile and they were generally of pretty good quality."

September 20 was "spent in walking around the outcrops of the Coal City and Broken Arrow seams which have been very carefully traced out by Mr. Brewer [Alabama Geological Survey] by means of test pits and auger holes . . . I went very carefully with Mr. Brewer over his line of test pits on both seams . . . There appears to me to be no doubt about his identifications."

In Coal City on September 23, Smith wrote: "Spent the day in the Black Ankle country which begins about two miles south of here along the boundary fault . . . and extends south westward with an average width of something over a mile . . . The Black Ankle is correlated with the Chapman and Ragland seams and contains most excellent coal but is not thick enough to make it worth while to mine." Three days later they camped in a section house west of Milldale. Turning homeward by way of Johns they enjoyed the cool weather and the "first class" roads of Jefferson County "over which we make with ease five or six miles an hour, and this gets us over the country with speed." Next day nearing Tuscaloosa. Smith was surprised to find that "the old plank road grade down to Hurricane Creek on the west side has been reopened to the very great advantage of all users of the road. Now if the overseer would go back to this old grade wherever it has been left by preceding overseers we should have in Tuscaloosa at least one first class road."

1900

CHOCTAW, CLARKE, DALLAS, GREENE, HALE, MARENGO, PERRY, SUMTER, WASHINGTON, AND WILCOX COUNTIES

The field trip of 1900 highlighted water. Robert S. Hodges went along with Smith as well as his son Truman with Dudley driving the wagon. On July 12 in the vicinity of Eutaw, they took samples and made photographs of Williford's overflowing well that had been "bored fifty years or more ago by a negro man owned by Dr. Prince. It is 400 or 500 feet deep and gives a stream about an inch in diameter. The water is palatable with little or no taste of salt or sulphur." Smith remarked the changing topography here at the last outcropping of the Tuscaloosa formation, from steep hills to "more rolling country with fine oak, gum, and short leaf pine growth indicating good soil."

After leaving Eutaw, they met with trouble at the ferry on the Black Warrior River—the road was almost blocked by large deposits of sand that had washed

Smith's emphasis during the 1900 summer tour was water wells. This photograph shows the Banks Well in Greene County.

into it. "We had to shovel it off from one side of the road to the other before attempting the ascent." On the Millwood Road, they passed "the handsome Cocke place, ... and also the old Cocke place with its houses with broken roofs. As we stopped near here for dinner, Mr. Hodges got a photo of this rather picturesque old house." Just before reaching Forkland, they saw "a large rectangular Indian Mound twenty feet or more in height about 150 feet long by 95 or 100 on top where a small patch of cotton is now growing." Rolling along toward their campground near Tishabe, the wagon fell into a hole, three feet wide and three feet deep, coming out of it with a bent axle.

Passing a field, Smith spotted a pine with "a curious deformity ... bulging rings around the tree." A nearby workman agreed to "cut it down and ship a 12 foot length of it to me which he says he can do for $3." On the roadside approaching Gainesville, where General Nathan Bedford Forrest and his men laid down what arms they had left in April 1865, Smith noticed a large cannon almost buried in thick layers of sand and pebbles. "It would be a good thing to have this on our University campus," he wrote. He met several men in town, from one of them he "got the fragment of mammoth tooth taken some years ago out of the bed of Factory Creek." A mile from Warsaw, he visited with Mr. White, an experienced well-borer, who promised "to copy off for me all of his records which have been kept accurately by him." They reached Epes just past six p.m. ". . . and after getting up forage camped on the hill back of the RR cut."

On July 24 at Livingston, Smith wrote: "Dr. Randall promises me an Indian pipe now in the hands of his married daughter. He also has part of a jaw bone with a large tooth, which from his description I imagine is part of the mammoth from the hole in Factory Creek. He could not find the thing as it has been lying about under the house uncared for, but says he will have it looked up if possible."

Next day on the way to York, Smith visited Alexander's Mill, "a rather large establishment with saw mill, planer, dry-kilns, etc." About noon at Millville, rain showers began, one of them "turned out to be a hard blowing rain which we had to take standing as it came down too hard for traveling." When it was possible they went on to York, but as the rain continued they spent the night at the hotel where the wagon, Mr. Hodges's buggy, and the mules could be sheltered in the barn.

Next morning they rode the mules out to Mr. Altman's to collect water samples from his well. "This water . . . has a very strong bitter taste like Epsom salts, and it is said to have the same effect when taken. The negros on the place drink this stuff regularly and profess to like it, though they say they cannot use it while they are working in the fields because of the weakening effects . . . We got half a dozen quart bottles of this water which I boxed and sent to Tuscaloosa."

From York they took the Cuba road on July 26: "The road had all been ploughed up yesterday and subjected to the rain that fell then. We got some pretty tough pulling out on these hills before we got to our lunch place . . . While here as a matter of course the usual rain came up and as it continued without any apparent intention of quitting we hitched up about three o'clock and came on. We found the road in the most execrable condition, the very next thing to impassible because of the recent working and the rains. At times on level ground in the flats, we could hardly pull through, the wheels sinking in to the hubs, and going up the hills in similar condition gave the mules all they could possibly do . . . We came in Cuba about 5:30 and met up with Mr. David Ward who allows us to camp in his enclosure under trees and in a very desirable place. We have a lot and stalls for the mules. Water (driven well) convenient."

At Gaston they photographed an impressive gully that had "cut back nearly to the road and in this we see the stratification plainly down about thirty to forty feet. Red loam ten to fifteen feet at surface, with white sands below . . . [that] are very sharp and are almost entirely fragments of white quartz, but in some of the layers there are rounded lumps of white clay an inch or so in diameter. This sand, if accessible, ought to be very valuable for building purposes."

As a result of the daily rains, the streams rose. When the expedition came to Double Creek, Truman tested the depth of the water by riding a mule across the stream. "The water came over the mule's back and we therefore concluded to retrace our steps and continue up the Demopolis road . . . Camped eleven miles from Demopolis. The camp was in the river flat, shallow dirty well, all the people in the house sick with chills and fever."

Smith wanted to go to Jefferson but learned that a slanting bridge made that road impassible for their vehicles. They "turned in through plantation roads and wound about through almost bottomless sloughs and over washed hillsides for two hours, and then back into this slanting bridge road about three miles from Jefferson." Arriving at noon in Jefferson, they stopped to eat at "a beautiful level spot for a camp under red oak trees with a fine pool of water just under the hill nearby. We all felt defrauded as we should have had this for a camp last night." Smith discovered that the creeks in the area were mislabeled on the map: "Coming southward across them the succession is Falling Creek, Double Creek, then Allen's Creek. These go together to form Double Creek below Jefferson."

On August 3, they stopped by Dayton long enough to take dinner and get their mail, then headed "down the Shiloh road which about four or five miles from town has been thrown out and a new road substituted. This to accommodate Mr. Bailey who has a sawmill by which he wanted the public road to pass. As is always the case with new roads, this was execrable, stumps left nearly as high as axles, bogs, washes, and every conceivable disadvantage which a road could have. From Bailey's place we did not follow the new road but turned down by Mr. Tom Boozer's into the old Shiloh road. Near Mr. Boozer's we passed a number of cuts in the road where the ledge of shell casts shows up . . . In Mr. Boozer's enclosure these shells are seen strewn along the hill sides and in washes. There must certainly be a quantity of these phosphatized shell casts sufficiently great to be utilized, by the farmers at least. If they were pounded or ground up and put upon the land in this fairly divided form they should be equal to 'floats' in fertilizing value."

Truman came down with a fever that caused Smith to change his plans and return to Thomaston for a doctor. They camped just outside of town under "the only shade tree available." On the third day, with Truman's fever still high, "we packed up making the wagon as comfortable as possible for Truman, and drove up [to Uniontown] by noon. After dinner Hodges and Dudley went towards Tuscaloosa [with the wagon] while Truman and myself will go later via York to Tuscaloosa. Thus all my plans are changed for I fear Truman is in for a long spell of fever." Smith, nevertheless, collected data about local artesian wells from a former mayor of Uniontown before taking Truman on to York where the night of Monday, August 6, soon after going to bed, "hot with fever," the boy broke into "a profuse sweat," and the fever vanished.

Smith sent word to the wagon to return, which it did on Wednesday the eighth, and they stayed around York "the rest of the week to recuperate." However, on the following Sunday, August 12, "Mr. Hodges began to complain and he was in for it regularly on Monday," Smith recorded. "This delayed us again till today [August 15] . . . but we came on without Mr. Hodges who is to join us at Akron tomorrow evening if he is well enough. . . . We camped under the pines back of Mr. Mosely's store [near Hulls on the Greensboro road], the only drawback being the presence of hogs, which, however did not trouble us too much around the wagon as they found enough to keep them interested about the mules."

The next day, Smith drove the buggy to Akron to meet the train. He picked up their mail, but Hodges did not come. By August 18, they were just north of Linden where they camped "in Mr. Jackson's pasture for want of a better place for there is a dearth of trees except immediately in front of the houses and on the edge of the creek swamp. We fear mosquitoes which have showed ample signs." Road conditions strained Smith's descriptive abilities—washes, gullies, and "a mud bed of terrible nature."

Moving on toward Butler, Smith and Truman paused to collect shells. "This we found to be most tremendously hot work and when we got back to camp I had to change every thread of my clothes as there was not a dry spot on anything." The family they stayed with in Butler had members sick with fever: "Dr. Moody says he has never known so much sickness in the county before. At Ararat the report is that 20 died during one week."

Near Butler, they visited a turpentine camp where they "witnessed the drawing off of the boiling rosin and the recharging of the still . . . The rosin was strained first through a wire screen and then through cotton batting spread over the wire screen bottom of a rectangular strainer and into a large rectangular tank from which, after some cooling it is dipped up and poured into barrels." After leaving the camp, they lost an hour or two by taking the wrong road.

On the way to Rescueville on September 24, Smith noted the tree "where the mob hanged a lot of Simses and Slaughters who shot so many people at the McMillan place." He was referring to Christmas 1891 when Bob Sims, believing that Mr. McMillan, the store keeper, had informed the law about his whiskey still, led his gang to surround the McMillan house and set fire to it. As the McMillan family fled the burning house, the Sims gang shot them one by one. According to James F. Sulzby Jr., "Only one survivor, a two year old baby, was found in the yard clinging to his dead sister." Later Bob Sims surrendered to the Mobile militia with the provision that the Sims women and children not be harmed. The enraged citizens insisted that the surviving Sims men be hanged, all six of them at the same time from the same tree. Then the Sims house and everything in it was burned.

Along the way near Shubuta, "in prairie country," the expedition visited with the Trice family, including former university students, Dan and Peter, who were having an artesian well bored. Dan guided Smith's group to "the fields in which the Jackson shells, etc., including Zeuglodon bones occur. I got here a number of pectens (with ribs), a few spines and vertebrae of fishes and saw several vertebrae of Zueglodon badly preserved. A Smithsonian agent has lately gotten two nearly perfect skeletons from a locality near P.O. Cocoa

Salt well, Cullom Springs.

a mile or two from Trice's on the road to Melvin. We stopped for lunch near Mr. Utsey's who lives a hundred yards from the McMillan store and house which was the scene of the murders by the Sims crowd."

The geologists ended the day camping by J. L. Pinkerton, at Rescueville. Pinkerton, Smith wrote, "is engaged in making an extract from some pyritous shale associated with a lignitic bed near his house. This he calls Pinkerton's Mineral Extract. He is also greatly interested in some iron ore which crops out around the hills at a certain distance below the top, just over the white lime rock that makes the base of the hills. This limonite is rather ochreous as a rule but there are streaks through it all of very good compact ore. We spent this morning [September 25] in tracing this outcrop around the hills and viewing the ore thrown out of dozens of pits along the branches flowing down the hill. We walked some five or six miles as the outcrops follow the contours of the hills around a space about three inches by two and a half, and the aggregate of ore to be seen is very considerable even if it does not extend back from surface into the hill more than a few feet. As I have said the general run of the ore seems to be very good and I hope it will turn out all right for Mr. P.

"After dinner we went down one of the ravines below the ore outcrop and saw on the white limestone a good deal of the fossil resin. . . . Below this comes a greensand shell marl . . . and just under that a black clay more or less lignitic in streaks. This is the material, especially the part most highly charged with organic matter (and incidentally with pyrite) that is dug up, spread under shelter till oxidation of the pyrite into sulphate takes place, then placed in hoppers and leached out with water, yielding the mineral extract which cures all ills. Much of this lignitic clay is full of leaf impressions. . . . In this layer there are many trunks and branches of trees, carbonized. Mr. P. sent us off with a bottle each of his extract and we have been well treated by him."

From Rescueville to Bladon Springs, Smith admired "the beautiful greensand . . . in all the cuts along the road." On Sunday, August 26, Smith and Truman rode out a mile from camp to take a guided tour of the various waters at Cullom Springs—Sulphur, Vichy, and Ferruginous. "The main point of interest," at least to Smith, "was the Artesian well which gives a very large stream of 83° temp. The water is accompanied by a quantity of gas which collects in bubbles on the surface of the tank into which the water pours. A lighted match touched to this foam ignites the gas which burns over the surface for several minutes before it is all consumed and in fact until extinguished, for a new supply is poured in without interruption. Also at the spout the gas may be ignited, where it burns with flame six or eight inches high. The water is very salty and a bath, taken under the spout, is a thing to be remembered. I have arranged . . . to get a doz. bottles of this water for analysis this winter."

From Robert Hearin, Smith learned that this well was bored "about fifteen years ago—water was reached at 80 feet but did not run over except in very small quantity. At 200 feet there was a bold stream of mineral (Vichy) water which over flowed. At 400 another strong stream of mineral water similar to that reached at 200 feet. Then at 1000 ft the stream of salt water and gas was struck. The boring was continued down to 1345 ft. but below 1000 ft. the pipe was withdrawn and the hole was filled up. The present flowing stream is the combined flow of 200 – 400 – and 1000 feet. The gas was used at one

Smith appreciated the occasional well-kept and prosperous homesteads that he saw on his tours, such as the Granade home near Frankville.

time to heat the bath water [at the hotel] and could easily be used in the same way now."

South of Bladon Springs, the saltwater wells at Frankville supplied the Confederates with salt during the war. Mr. Granade told Smith many of the wells in low places were only eight to twenty feet deep. Smith wrote: "The salt often crystallizes out on the flats in dry weather. Besides these surface wells, others were bored as deep as perhaps 400 feet yielding also salty water." About nine miles from Frankville, they passed a large oak on the edge of the road, a place, Smith remembered, where he and Langdon had camped fifteen years ago.

August 28 found them at Old St. Stephens. Smith and Truman took the buggy, and with a local guide went sightseeing—to the old fort on the Tombigbee River bluff, down the main street of St. Stephens, then to Caroline Medora Brown's memorial, a tall obelisk adorned with a wreath of stone roses and an elaborate inscription. Mrs. Brown died in childbirth on April 29, 1841, nine days after her eighteenth birthday. Smith took photographs at all three of these significant places.

Upon returning to camp they hitched up and headed for Jackson with Smith admiring the flat, level, beautiful farming lands along the way. They had a hard time finding a place to camp in Jackson, finally setting up on a red clay hill behind the cotton gin. The next day Smith spent the rainy morning writing about thirty letters. Late in the afternoon, he and Truman "went down to the spring station where they have piped up the sulphur water and delivered it through the bottom of a very nice marble vase." After supper Daniel Langdon arrived unexpectedly by train. The group spent the following five days collecting on the river bluffs but found nothing outstanding. On September 4, Langdon set off for Tuscaloosa leaving Smith to pack and nail up a box of sand and shells to send to him.

As they broke camp on Bell's Landing Road, "Dudley began to experience the pains of a chill and was down with fever in a short while so that Truman drove the wagon up to [Lower Peach Tree]. The rest of the day after 11 o'clock was spent in our camp . . . near the warehouse with Dudley lying up with fever. We took our supper with Mrs. Stallworth whose son keeps the warehouse. Before leaving [September 7] and while waiting past Dudley's chill time we went across the river and took pictures of the bluff. . . . We sent Dudley on with the wagon to set [up] the camp . . . and when we overtook it, we found him lying up with another chill only 36 hours from the first which was contrary to the 24 or 48 hour rule for such things. We had therefore to cook supper for ourselves which we did with fair success, notwithstanding a rain."

Two days later Smith wrote: "We had to cook for ourselves again this a.m. as Dudley was still quite unwell. I staid with him last night to give medicine. At the ferry [on the Alabama River] we were congratulating ourselves on the large flat with stout railings, but we got aground several feet from the hither bank and had much trouble in rolling the vehicles by hand over planks to shore where the mules took them. Camped in Mr. Rush's crib [at Prairie Bluff], but we take our meals in the house. Heavy rain and bad roads and all sorts of trouble."

On September 9 at Glastonburg, they set up at a good camp and had a hospitable host. Smith went with Mem C. Weir to collect shell casts on his plantation and "one large hollow cylinder which looks much like a Hippurite but different from any yet seen by me," Smith wrote in his journal. He added: "Dudley misses his chill for which we are all thankful."

Orrville surprised Smith on the tenth by its prosperity—"a number of nice new houses," "a first rate store" along with several average ones, and a large cotton gin. Traveling over the Eutaw sands and the Tuscaloosa formation on September 12, Smith commented on their striking differences: the Eutaw characterized by fertile soils and hardwood trees, the Tuscaloosa by "poor pine hills, the farms showing in places almost sterility." At Mr. Thomas's place where they stopped for lunch, Hodges inadvertently dropped two $2 bills. The geologists had progressed nearly a mile on their way when Thomas caught up with them to return the money.

Smith had scathing words for their route that day: "The posting along this road is something to wonder at. Thus we passed 37 mile post from Tuscaloosa three times in a distance of about three miles and coming on some three miles further we pass the 38 mile post. . . . Taken altogether the road from Marion to Tuscaloosa is one of the dreariest, barrenest, and in every way worst that I have yet encountered, [one hill after another covered with fragments of iron crusts, with deep sand beds between] excepting of course the unworked roads of Marengo."

1901–02

BALDWIN, CHOCTAW, CLARKE, COVINGTON, ESCAMBIA, GREENE, MOBILE, SUMTER, TUSCALOOSA, AND WASHINGTON COUNTIES

Dr. Charles Mohr, who spent nearly forty years of his life in Alabama, died July 17, 1901, in Asheville, North Carolina, where he had moved for health reasons. His *Plant Life in Alabama,* to which Smith contributed information and specimens, came out shortly after his death. The book, consisting of 921 pages and several plates, was termed by Dr. Roland Harper "the best description yet published of the vegetation of any whole state or similar area."

At the time of Mohr's death, Smith was in Mississippi touring water wells in Artesia and Aberdeen. In Alabama again August 29, the water wells of the St. Stephens area had his attention. "At one of the wells there is the story that when they were boring it there was a great explosion or blow out of gas which tore up the tube, making a loud noise heard for miles. With this explosion came a great rush of gas which caught fire from a camp fire some forty or fifty feet away and struck back to the well and was communicated to the derrick destroying it. This story is repeated by quite a number of witnesses but whether or not much exaggerated, who can say."

On his way to the ferry on September 1, Smith noticed an exposure of ferruginous sand having among its shells a coral thought by most of the locals to be bones. About "Bucksnort," a productive salt well near the river, Smith wrote: "In one of the pipes there can be heard a great frying and bubbling of gas and this may be ignited at the top of the tube whence it will burn back with a whistling noise like that heard when a half explosive mixture of hydrogen and air is burned in a test tube."

On February 3, 1902, Smith was in Washington, D.C., testifying before the mines and mining committee of the United States House of Representatives. At stake was a fifteen thousand dollar appropriation for each state to set up mining and mineralogy schools. Smith was there to get the money for the University of Alabama, Professor Thach of Auburn had come to take it home to Auburn. By February 11, the two men, working with Alabama congressmen, had devised a plan to divide the fund equally between the two schools.

In mid-June of 1902, Smith escorted his daughter Julia to Montgomery where a crowd of Smiths and Pratts gathered to see her and Professor Smith's sister, Julia Pratt, off by train to New York, there to board a steamship for an extended stay in Europe. By June 21, Smith had returned and was out in the field, meeting with Mr. Thompson at St. Stephens concerning the oil well he had drilled, and the larger one he

Merrill, Smith's youngest son, mounted on Old John at the survey's campsite in St. Stephens, once the territorial capital of Alabama. Probably photographed in the summer of 1902.

The survey crew enjoyed the health springs where they mixed business with pleasure. Smith collected samples from the waters for later analysis. This photograph of Borden-Wheeler Springs, Cleburne County, captures the relaxed atmosphere of those popular health resorts. Photo by Smith's assistant, Robert Hodges.

anticipated drilling with another kind of machine. In the same area, Smith visited a generous source of excellent limonite, but "probably not enough to be made the basis of any extended mining operations," he concluded.

South of Bladon Springs, after the last lime hill, they crossed Shell Branch. Here they walked upstream collecting many shells. They spent the afternoon and night at the Cullom Springs Hotel, which, Smith noted, was "newer and more comfortable" than the hotel at Bladon Springs. The recently refurbished hotel, in addition to the usual resort amenities, had electric lights and fans, long distance telephone connections, daily mail service, and a large swimming pool with a continuous, one hundred-gallon a minute flow of salt water.

However Bladon Springs was still considered one of the premier resorts of the state, and its Old Bladon Spring Water was sold far and wide. Situated in an oak-hickory-pine valley three miles from the Tombigbee River, the hotel was about twelve steamboat hours from Mobile. It was frequented by famous personages from the world of literature, music, and politics. Though near neighbors, the two resorts flourished and were booked up every summer.

On June 23, the wagon crossed into Mississippi where near Bucatunna the geologists visited "Mr. Conner's Natona earth establishment formerly the Acid Iron Earth." As did Mr. Pinkerton in Alabama, Conner dug the pyritous clay, spread it under a shed to oxidize, concocted an extract, but then, unlike Pinkerton, he manipulated the result "into tablets . . . which can be dissolved at will and used." They stopped for dinner at Healing Springs where the proprietor, Mr. Pettus, showed them a bog from which seventeen springs rose several feet above the bog's level. Smith agreed to analyze the four springs that furnished the water, noted for its clarity and pleasant taste, for Pettus's hotel. Healing Springs differed from most of the current resorts—it stressed simplicity and healing rather than fashion. Recently a Mobile minister who experienced healing at the springs had opened a school that allowed students to work on the school's farm to help earn their way. At that time the Healing Springs Academy was the only accredited high school in Washington County.

After breakfast in Citronelle on June 24, they spent half a day inspecting "the Derrick and the oil seeps, which were not active . . . today possibly because of the dry weather." They headed north into Mississippi again and spent time in Enterprise observing Mr. Page who was boiling down some material on which the scum "rises to the surface of the water [and] . . . has a very strong smell and taste of kerosene," Smith wrote. "I also saw a good deal of oil film in the water at the seep." On to Quitman where they took dinner at the Heim Hotel run by a German woman who no doubt reminded Smith of his memorable student days abroad. Then to Shubuta where he took a long walk to inspect the bluffs of the Chickasawhay River. All of this area of Mississippi had suffered from recent flooding of the numerous creeks and rivers—houses and farms had washed away, the shells Smith searched for had been carried off, deep sand drifts lay everywhere. He came upon a crew raising "a new steel bridge and repairing the embankments washed out last spring." They gave him a ride back to town in the engine of their construction train.

Returned to Tuscaloosa by June 29, Smith sent two packs of paper to Truman, now at work in Gaylesville, "one quite thin for writing to Julia [in Europe] and the other thicker for ordinary writing." In a July 11 letter to Truman, Smith complained that the family still had not heard from Julia, and that they were "getting to be a little impatient." He wrote of his intention to send ten-year old Merrill, the youngest son, with the dog Bob off on the wagon with Mr. Hodges next day so they could have a day on the road before he joined them. He wanted to see how Merrill would get along without him in case he had to leave the youngster with the wagon sometime. He closed the letter to Truman with the admonition to cut back on his hayrides and other social activities before they interfered with his work.

To Truman again on July 12: "Well, I got the wagon and buggy off this afternoon about half past three and I suppose they will go tonight about as far as we did in 1900, that is two or three miles beyond Foster's Bridge. Tomorrow they will go on down to Eutaw and I will catch up with them there tomorrow night by the train. Merrill cried a little when he went off and talked a little homesick, so I should not be surprised if I had to bring him back Monday. Still, after he gets started he may get over it. He took Bob along. The new tent is a daisy, it has an awning in front under which we can sit and take our meals and it will be fine in a rain. Merrill and Thomas Palmer and Charlie slept on cots in the tent last night and we gave them a good supper under the awning, in which all of us joined. They were awake and playing about this morning a little after day, and all said they had a fine time of it . . . No word yet from Julia, though we are looking for a letter by every mail. You know the steamers sail only once or twice a week, and that may cause a good deal of delay, especially if they did not get letters off immediately on landing. Your mama is feeling mighty lonely now without Merrill . . . Jeff [the dog] looks lonesome too."

The wagon with Jefferson Davis Jackson driving had left Tuscaloosa on Saturday afternoon, reaching Sylvan by camping time. On Sunday morning they rolled on to Eutaw to meet Smith arriving on the evening train. Monday night, three or four miles below Epes, they camped at Mr. Henegan's, "going

Smith's wagon with patient mules attached and Robert Hodges's buggy at a ferry crossing in 1902. Merrill, youngest of Smith's sons, is perched on the rail.

into the house (vacant except one or two rooms occupied by the servants in charge)." On the way there, they had trouble crossing the Black Warrior—they waited an hour for the ferryman. Finally his brother in law "set" them over. A violent storm during the night—"a good deal of wind accompanied by much thunder and lightning"—made them "glad to be in a house instead of camp."

By July 16, the wagon crossed into Mississippi where "after making a few purchases in [Meridian] we drove out the Enterprise road over Natchetigbee to the foot of the 'mountain' and some distance up . . . Camped near the first summit of the mountain in a place where the spring is difficult of access and not very prodigal of water. Merrill is not feeling well."

The next day at Enterprise, Judge Johnson and Aldrich joined them. Merrill had "a little fever" that Smith wrote, "throws me all out of calculations." The rest of the party took the scheduled trip to Watubbee, but Smith instead conferred with a doctor about Merrill who had a temperature of 102 degrees at noon. However by four p.m., it was dropping toward normal, and Smith was able to join the next day's venture. They hired a rig for a long day's side trip—they did not get back until seven that night. They rendezvoused with the wagon near Ben Donald's place where Jeff had set up on a west-facing bluff. Smith found the site "intolerably hot" in the sun. "Mr. Aldrich and I, after the sun was off, took a fine bath in the river." As they pulled out of camp next day, "the mules broke the pole of the wagon which caused a delay of some hours."

The geology wagon was back in Alabama by July 22 in the vicinity of Bladon and Cullom springs. They made lunch at Frankville. "Zeuglodon bones have been found in all the prairie fields about here. Very few are seen now." At Bigbee they took several hours to see the turpentine still and barrel factory of Boykin and Boswell. "Today [July 23] we have come about 26 or 27 miles down by McIntosh Bluff," Smith recorded. ". . . The level lands are beautiful farming lands and we passed several beautiful old places with rows of great water oaks on each side of the road before the houses. Wilsons and Nick Harris are two of the places. Copelands Creek or the one into which it flows is a beautiful clear stream in which from the bridge we saw several large fish a foot or more long."

After McIntosh Bluff, they went south to Malcolm where a tempestuous rain and wind tested their new tent. They visited Dr. James T. Searcy, who had succeeded Dr. Peter Bryce as head of the State Insane Asylum in Tuscaloosa. He was also in charge of the newly established branch hospital for the "colored insane" at Mt. Vernon in Washington County where Smith's party was put up in "good comfortable quarters." Smith and Dr. Searcy, on a Sunday drive, toured the Mount Vernon landing, the Hardwood Sawmill, and the Indian village or, as Smith explained in his journal, the site of the village "as most of the houses have been torn down."

After reaching the coast early in August, the mosquitoes and sand flies gave them no rest especially when they camped at Bayou La Batre. Smith was interested to learn that Mr. Bromberg's well at the edge of the salt marsh produced superior water for the region and was delivered every day to the coastal rooming houses. "From Bayou La Batre we drove down to the oyster canning factory and along the coast by Sans Lovice Park, just now vacated by the soldiers, and up here [Coden]. We had the good fortune to get a camp in the grounds of Capt. Alba by whom we are most hospitably treated." Smith visited a brick manufacturing plant near Coden that "uses a pug mill and the material is obtained in the immediate vicinity. The clay used . . . is a very sandy clay of yellowish color." The journal entry for August 2, 1902, continued: "Yesterday and the preceding day were characterized by very high winds from the south. Clouds were numerous and last night they thickened up and the wind rose almost to a gale. After we had retired the rain came and the tent pegs no longer held and we had a time of it trying to keep from being blown away."

That same day they returned to Bayou La Batre by way of the beach to photograph the artesian well tank and the canning factory. "The well is 580 feet deep through sands and clays. The first water was reached at 500 feet but the yield was small, 100 gallons per minute and the water rose only to the top of the tube. The boring was then continued forty feet when there was a sudden drop of the tool as through a hollow of twenty feet. The water was reached again at the depth above mentioned and it rises about fifteen feet above the ground. The tank is filled direct, and from it a wooden trough about one fourth mile or longer carries it down to the factory."

On August 4, they "put up at the hotel of one of the younger Delchamps. The boring for oil is going on near the saloon and they have reached a depth of seven or eight hundred feet. [Today] . . . we hired Mr. John Reed's sloop and went across to Fort Morgan and returning passed near Fort Gaines, the coast of Dauphin Island, Cedar Point and back again after a delightful day, good breeze, fine weather, and everything to make it pleasant. After returning four p.m. we hitched up and came five or six miles up to this Creole settlement [Mon Louis] where we camped."

They were in Mobile the night of August 5. "The trip today has been from Mon Louis up the Bay road here, when we arrived about four p.m. making excellent time on account of good roads (shell road) for seven or eight of the miles. The Creoles have a ship yard as they call it on Fowl River and they are building several good sized schooners at this time." Rain the next day prevented them from packing up their wet tent and so they "concluded to wait over till tomorrow. The day has been spent in seeing different people and making purchases. This afternoon I went with Capt. Alba out to the place where the drilling of the Mobile Oil Co. has been going on. They have reached the depth of 1558 feet and have struck a great stream of salt water and gas. The water spouts up through a four inch pipe to a height of six or seven feet above the pipe. So much gas comes up with the water that it can be ignited by thrusting a burning paper into the stream. The gas burns then at the top, along the sides, and also apparently in the midst of the stream. A grand sight. Mr. Jovillian gave me some shells which came from a depth of about 1537 feet."

They crossed Mobile Bay to the eastern shore on the small boat, *James A. Carney*, with "the mules and wagons . . . packed in like sardines and there was very great danger of the mules becoming entangled in the wheels of the buggy or of kicking it to pieces. We got over all right, not withstanding. [With William Yuille] . . . I took a long walk up the Bay Shore beyond The Village to see, among other things, the great live oaks one of which is called the Jackson Oak, and the immense grapevine with trunk 18 inches to 24 inches in diameter. The walk up Bay Shore . . . is a very attractive one. We pass a number of nice houses with well kept yards, but some with yards neglected. We put our camp in the magnolia grove of Mr. Webb where we had a good place and good water."

At Montrose, with the help of vacationing Sam Hodges, chief cook at the fabled Battle House in Mobile, they "went down to the beach and to the Red Bluffs" to make several "good photographic views of the Bluff." Sam was so polite and helpful in every way, putting his boat at their disposal and carrying the heavy camera and equipment, that Smith promised to send him prints of the photographs. Five miles farther south they lunched at Fairhope, "the Henry George Colony of Western people." In the high flat piny woods land below Point Clear, Smith saw the peculiar sinks, "holding water usually but mostly dry. These are very numerous as we passed four or five within distance of two miles. They vary in size, some occupy about an acre, others are much larger. Near our camp one small pond supplies the water for our mules and the cattle about here."

At the head of Week's Bay on August 9, they stopped for several hours to tour Mr. Nelson's turpentine distillery where they were treated to "some fish from the bay. Mr. Hodges took two views of the distillery and wharf . . . After dinner we started to come up . . . [to Magnolia Springs] and then go further south, but as we approached the ferry over Fish River it was over a road built up of muck from the marsh. This was about 100–150 yards long, and quite narrow, and what is more to the point very boggy except for a thin dry crust on top. Soon after we struck this, the hame string of one of the wagon mules broke and the wagon got a few inches off the beaten track, and the wheels on that side sunk in eight or ten inches. The mule John on that side began to bog up and for a while it looked as though wagon tongue and a number of other things would go by the board. The hook from one of the single trees broke out and we had to make a contrivance with wire which Mr. Moore who keeps the ferry, gave us. I could not risk hitching up mules again on this bog so we pried the wagon out of the muck and then pulled it by hand to the ferry . . . We got here [Magnolia Springs] about seven as we lost an hour or two in the bog."

Camped eight miles from Bay Minette August 14, Smith wrote: "Yesterday morning we started out by the Bishop's Bridge [over Fish River] road and came in to Daphne about 4. [On the way] . . . we passed through a considerable tract of virgin pine forest which was a fine sight. The ground covered with beautiful green grass, without a trace of Undergrowth or shrubbery, except where gums and titi were seen around the margin of a pond. In one place there was a long curved space without pines looking exactly as though it were a filled up channel of a winding stream. We took a photograph of some of these trees. Beyond this we came into a Swede settlement or a Company grounds which lasted almost to Baldwin's Store on Fish River at least four miles. Along this land were houses and fields and peach orchards, grape vines, etc. Going down towards the store we got of course into sands and washes and live oak thickets. From Baldwin's our route was up the river two miles to Bishop's Bridge which we crossed and on the hill beyond at Mr. Bishop's we stopped for lunch. Mrs. Mary Bishop here, Marlowe P.O., was very urgent that I should go and see her son by a former marriage John Rollins in the Insane Hospital at Tuscaloosa and write her about him, which I agreed to do . . . [Mr. Webb] spoke of shallow wells in the strip of land or sand between the Little Lagoon and Gulf as yielding water more highly charged with salt than the Gulf . . . From these brines salt was made during the war."

On Saturday, August 16, 1902, they "got to Stockton about 12 and here a meeting of Confederate veterans was in progress and several hundred persons in attendance. Mr. Feminear invited us to come to the table for dinner which we did, partaking of the good things there provided chief among which to my taste was the roast beef." Next day in Bay Minette they "camped on the premises of Mr. Feminear about three fourths mile down the RR and waited all day for mail which did not come, for some reason as yet unexplained."

On August 18 in Clarke County, the geological expedition spent an hour at the Gainestown landing on the Alabama River "shouting for the ferryman. When he came we went over in a skiff and I had to walk a mile to find some-one . . . who could give me some information about the marble. As I came back I was overtaken near the ferry by Mr. Flinn, son of Mr. Robt. Flinn of my time at the University and the regular ferryman, from whom I got as much information as was needed. Going up on the terrace from the river flat the road follows a tremendous gully along sides of which the following section is shown" and Smith sketched the strata (five of them) in his journal.

"Some of the strata of this limestone is calcitized and these constitute the marble," Smith continued. "The large blocks of stone just below the ferry road are of this rock. They were gotten out during the war for the purpose of obstructing navigation. At Maj. Marriotte's [where they were at present camping] I got two polished specimens one yellowish in color, the other red, which show up very well and if any large masses of this variety can be had the rock may certainly be put to ornamental uses. It seems on first inspection that the difficulty will be to get large enough blocks of sufficiently uniform quality to be of use."

Near Douglasville, August 23, 1902: "Mr. Wyatt is making an extract of the pyritous lignite at Bluff #3 which he uses for medicinal purposes. Some of this submitted to State Chemist has been analyzed . . . showing a considerable percentage of nitrogen and Mr. Wyatt has hopes of inducing some fertilizer firm to take hold of it for him. I told him I thought the chances very small of finding the material of sufficient uniformity of composition to make the working of it a safe venture. At Roberts also Mr. Murphy at the store showed me some of the iron sulphates which he had sent to me, but not being satisfied with my report, has just sent some off to Auburn in hope of getting some report more in accordance with his wishes."

On August 26, they went with Mr. Dixon to his mill to see the strata of exposed white limestone. He promised to send Smith samples of clay that in his opinion were worth examining. "On our tour we got some fine watermelons and grapes. As we came back Mr. Hodges took two pictures of the Blue Pond which is just over the line in Covington [County]. While there we heard the whistle and other noises of the log train, and so we turned back and got two views of the log train and engine, a curiosity in its way. We did not get back to camp till about 12."

As they wanted to move on to Andalusia, they "took a biscuit or two for . . . lunch and got under way about 12:30 and came into Andalusia about 6, a drive of 18 miles . . . After getting mail and feed we camped in the grove just by side of the grave yard, a first rate place for camp barring the millions of black ants which covered the ground everywhere. Mr. Rankin, a former student at the University, met us and was very helpful in showing us the camp and in several ways saving us time. During the night, a number of messengers brought in news of the primary election which was held on this day."

They reached Montgomery on August 29, but heavy rains and dark clouds toward Prattville, their destination, made them hesitate. However, they decided to forge ahead, pushing the mules to make twenty-nine miles that day before camping in a vacant store. "Soon after stopping Old John Mule developed a severe case of colic, caused us for a time to count him as a dead mule but pepper tea, paregoric, whiskey and soda finally prevailed and I lay down between 2 and 3 a.m." The next morning they arrived at Prattville "before dinner time, stopping a few minutes to take a photo of Buzzard Bluff and one of the Ferry." They stayed at Prattville visiting with Smith's relatives, "writing letters and looking about" until Mr. Hodges departed for Tuscaloosa on September 1 and Smith set out for Texas the following day.

October brought sorrow to Eugene and Jennie. Their twenty-nine year old first born, Eugene Allen Jr., who seemed to be especially close to Jennie, died on the twelfth. His was the first grave in the Smith plot in Evergreen, the cemetery near the university. The *Crimson-White* wrote that his death "came as a shock to all, for it was not generally known that he was in failing health . . . Quiet and unobtrusive in life, faithful in the discharge of duty, of the strictest integrity, he was the possessor of warm friends." The funeral was held in the Smith home on October 15 at three-thirty o'clock. During the service, a quartette from Christ Episcopal Church sang "Asleep in the Arms of Jesus." At the cemetery, the floral tributes "completely hid the grave."

In November Smith made a short foray to the Spring Hill section of Mobile to observe two gas wells—one producing about two hundred thousand cubic feet of gas in twenty-four hours, the other in the process of being drilled. Mr. Jovillian, whom Smith and Captain Alba saw in August, had saved "a very fine lot of shells" for Smith, and he gathered even more "from the sands which were spread out on the flat."

1903

Barbour, Clarke, Conecuh, Dallas, Jefferson, Marengo, Monroe, Sumter, Tuscaloosa, and Washington Counties

The *Tuscaloosa Times-Gazette* described an April "plant" party celebrating the anniversary of the discovery of *Neviusia alabamensis* (Alabama snow wreath) by the Reverend Mr. Reuben D. Nevius, who had been a long time rector of Christ Episcopal Church and a botanizing friend of Smith. Former students and old friends of Dr. Nevius including Eugene, Julia, Truman, and Merrill, gathered on a high sandstone cliff at the confluence of North River with the Black Warrior where the snow wreath was "in full bloom [forming] a beautiful fringe for fifty yards around the point covering the rocky cliffs over a space of twenty feet vertically," according to the newspaper. They picnicked beside a waterfall and afterwards entertained themselves with "Hunting, Fishing, Shouting." Before returning to Tuscaloosa by the M & O railroad, the celebrants visited the Holton Furnace Pipe Works, the coke ovens, and a product plant.

In June, Smith was off again to Mississippi. He especially looked for "the leaf bed and palmetto trunks" that Dr. Hilgard once found along the Chickasawhay River. At Bucatunna, however, he recognized that changes in the river channel and development of agricultural fields had obliterated the site. He collected at several other locations along the river but soon returned home.

With the Fourth of July approaching, the Smiths made plans for a gala celebration on their houseboat tied up at the mouth of North River. Smith wrote seventeen-year old Truman who was working at the Kellerman Mines in Tuscaloosa County, listing some of the guests expected, and urging him to join them. Truman did come, but he rode horseback from Kellerman—about fifty-four miles round trip—which worried Smith considerably. Afterward he wrote Truman: "I have been concerned about you all day for it has been so hot I am afraid the long ride has been too much for you. Write me at least a postal as soon as you get this and let me know how you [stood] the trip."

Jennie with Julia and the two younger boys spent a great deal of their summer time on the houseboat entertaining. Julia's out of town friends as well as "parties from town [were] continually driving or rowing up to spend a moonlight evening," the local paper reported. "They are anchored close enough to the bank to go ashore whenever they please and they are having a most delightful time, rowing, swimming, and enjoying life on the water and in the open air."

Smith and Mr. Hodges took a short trip to Epes. The postmaster had hired for them a man with a skiff that turned out to be, in Smith's words, "a heavy old row boat without oars, but with three paddles." They then embarked on the Black Warrior River for "thirteen hours of continuous paddling." Smith made little comment about that day's work, but the following day was less strenuous and more interesting. They hired a surrey to go to Hatch's Bluff. Along the way, they passed the imposing antebellum Hatch house on a bluff above the Black Warrior. On the return trip, they stopped to explore and photograph the now vacant mansion. Smith described it as "a brick building with columns and a marble tessellated floor in front, a porch on side and back. Covered way leading from house to a building [in] back which was probably used as a dining room."

The builder, Alfred Hatch, had grown cotton, corn, and sweet potatoes on his two thousand-acre plantation. Until 1861, when his wealth was estimated at five hundred thousand dollars, he was renowned for his annual Independence

Day picnics. After the war, he was worth only forty thousand dollars, and by 1870 his magnificent home on Arcola Bluff with its numerous outbuildings was sinking into ruin.

In mid-July, Smith went to Selma to evaluate the different cement deposits in the area. Nearby Orrville impressed him as being "one of the few old time towns that is doing well." He saw "quite a number of good houses, well repaired and well kept front yards with good grass sward and flowers."

On July 22 after a business trip to Birmingham, he took the train to the coalmining communities of Yolande and Searles, then walked to Kellerman to visit Truman. "After supper we went into the mines where one or two plugs and markers had to be placed. This mine is certainly the best worked of any I have seen in Alabama. All the walls are cut true and clean with no jagged edges and rough surfaces. They have many tons of coal piled around the entrance of the mines ready for shipment as soon as the railroad comes in. The track is at this time laid within four miles of the mine and the only delay will be caused by the trestles which will have to be put up over ravines." He inspected with interest the 1872 mine openings made by the Welshmen. When he returned to Tuscaloosa, Smith packed a box to send Truman. It contained, among other things, four novels, a baseball, and a football with Smith's meticulous directions for inflating it.

At the close of July, Smith went from Montgomery to Eufaula by train, making notes along the way from the rear platform, to meet with Roland M. Harper, mapmaker and botanist. In the afternoon, they walked ten miles out the Central Railroad then caught the train. Smith ended up in Clayton where he visited "the Devil's Elbow Cut about one mile from Depot towards Eufaula." He and his host went tramping about the countryside studying the kaolin and other clays that Smith thought merited industrial attention.

Meanwhile Jennie, Julia (in her mid-twenties and a graduate of Tuscaloosa Female College), and Merrill summered in Prattville with the numerous Smith relatives and friends. On August 5, Julia wrote her father their news in a cheerful, affectionate letter. She told him how Merrill and his boy cousins and their "little black chums" went swimming in the pond, and about her two offers of teaching positions, one in North Carolina, the other in Alabama. Cousin Dan Pratt had told her North Carolina was a better choice. "But that is so far away," Julia wrote her father. "I feel like I would a thousand, thousand times rather teach scrub-primarys at home than have more congenial work away from all those I care for on earth." She mentioned her friend, Mr. Yuille, as one of their visitors who had left. "I miss Mr. Yuille but on account of Mama's constant nagging and planning I am very much relieved to have him gone.

"I was awfully sorry to hear of Landon's trouble . . . Mama was terribly cut up about it at first, cried lots and seemed to think it meant permanent ruin and disgrace to Landon. But she is looking at it in a better light now and she told me to tell you she had written to Landon and said nothing about it. She got your letter and check yesterday . . ." Julia closed with: "I am so delighted to think that you are coming to see us soon. Your visits are the greatest pleasure." Later Julia followed her heart and chose to teach at the local high school.

At the time Julia was writing her father, he was in Jackson appraising an outcrop of orbitoidal chimney rock in the river bluffs—"quite compact and solid and pure. It takes very good polish and seems to be . . . pure carbonate of lime . . . Just north of the line of the State lands Mr. J. H. Hobson is quarrying the hard bed of this rock and shipping it by barges to Ft. Morgan for use in building a sea wall . . . Mr. Hobson showed me a polished piece of the marble from Salt Mountain. This is beautifully marked and is worth close investigation. A short distance above the quarry occurs the first shoal on this river [Tombigbee], so that from here to the Gulf nine foot of water can be counted on at lowest stage of the river. This gives Oven Bluff and the State lands an advantage over any place I have yet seen, i.e., if there is really deep water all the way down from here . . . On our way back we came by the Salt Works and then up Limestone creek to the sulphur spring. This spring furnishes a very pleasant water, salt and sulphur of about the same quality as that at the Jackson well."

On the next day, Smith was prowling at St. Stephens, taking clay samples, speculating about the lime rocks, hoping some of them could be combined to make cement. He concentrated on the area "just above the mouth of the ravine separating the old town from the bluff on which the Fort is located." He studied old chimneys made of the lime rock to see how the long years of fires had affected them. "So far as I could perceive it was just about like the fresh rock."

On August 8, Smith went into Jackson to pack and ship specimens then he headed for Suggsville. "For three or four miles the road was well nigh impass-

Left, Smith photographed the "beautiful little boat" under the Mobile and Ohio Railroad bridge on the Black Warrior River before leaving Tuscaloosa on August 27, 1903. Below, Naptha launch anchored at Elm's Bluff on the Alabama River. August 28, 1903.

able by reason of its being so cut up by the log wagons . . . All along the road traveled there are great beds of gravel . . . The hill here at Suggsville shows a world of gravel." The next day Smith visited his old friend and roommate, John Lee Marshall, not far from Perdue Hill, then on the eleventh went "to Monroeville with the mail driver. Double Branches which we crossed near town is a fine stream and should furnish a very good water power, it is worth measuring . . . I saw . . . a pretty good specimen of turtle back obtained from Curiosity Branch near Scotland P.O."

On August 12 in Evergreen, Smith visited a brickyard where the bricks were made of clay that came "from a sort of bog and is apparently of recent origin. It is claimed to be good for fire brick and these bricks as tested in linings around boiler fires have much outlasted brick bought for $30 . . . There is an artesian well here that supplies the town . . . It yields 130 gallons per minute and is pumped into stand pipe and distributed through the town." He continued in the Evergreen area on August 13 exploring quarries along Murder Creek before returning to town in time to take the 2:15 train for Prattville. He noted in his journal that he got "a rear seat in the car and so had a good post of observation."

Smith arrived in Prattville on August 14. Julia had very little time to visit with him, as he left the next day for Tuscaloosa. At home he found Truman confined to bed with fever that did not go down until the fifth day (August 19). He and Truman then rejoined the family in Prattville for three days, but Smith spent part of that time in Montgomery. He returned to Tuscaloosa to take care of mail and other pressing matters until August 27 when he left

Ferry and Englemann house at Old Cahaba. August 28, 1908.

White Bluff, photographed from the Naphtha launch during the 1903 trip.

with several others on a trip downriver in what he called "a beautiful little boat," a Naphtha launch. They left Tuscaloosa at 8:30 and were in Selma by midmorning. Seven miles below Selma they got "stuck on a sandbar . . . and were at least an hour getting off. We struck bottom a number of times . . . but had no serious delay except at this first bar.

"At old Cahaba I took a view of the ferry and the old brick house of Mr. Louis Englemann. We went up into the town and took a short look around. Quite a number of old time houses, some of them of brick, are still standing though mostly in a state of dilapidation. I saw two of the artesian wells, but not the main one which is some distance from the ferry and landing.

"From King's [Landing] down to White Bluff there is nothing remarkable to be recorded, but the approach to White Bluff from up stream is very imposing. This is the highest bluff by far in this part of the river, and it makes an unbroken wall, nearly vertical and almost 200 feet high, for a distance of nearly a mile and with some interruptions, caused by slips, which make a narrow bench in front of the main cliff, for two or three miles, perhaps more." He made three photographs, "a view of the upper end of White Bluff and a little further down, and the Bluff viewed from downstream." They ate their lunch "in a deep ravine or cut in the bluff where a branch comes down in wet weather."

In Selma on August 29, Smith "tried in vain to get a team to go down to Berlin and Pleasant Hill, but the weather was so hot the livery men were not willing to send their teams out, so I concluded to come out to Demopolis instead." Smith toured the cement works in Demopolis collecting specimens of limestone and clays. He closed the 1903 trip (and his third decade as state geologist) with: "After returning to Selma for dinner I took the 4:30 train and came to Tuscaloosa by ten."

In this ten-year period, Smith expressed his deep interest in the coastal plain of Alabama through bulletins, articles in scientific periodicals, and especially in the "Report on the Geology of the Coastal Plain of Alabama" for the Alabama Geological Survey (written with Judge Johnson, Langdon, Aldrich and K. M. Cunningham.) He also wrote reports on the gold ores of Alabama, the clay resources and their relationship to industry, the increasing importance of the state's cement resources, the Carboniferous fossils, the phosphates and marls, and other subjects. He spent a great deal of time and effort pushing the state to make the Geological Survey's discoveries available for use, to carry out what he considered the under-funded survey's purpose: "It outlines the geological formations with as much accuracy as possible, notes those in which ores occur, gives descriptions of some typical occurrences, and then every man exploits his own land."

For *Science* he wrote of his trip to the sulphur deposits of Texas. He wrote papers memorializing Michael Tuomey, Alabama's first state geologist, and his friend, Charles Mohr. He corrected and produced maps and reviewed reports of other geologists. He diligently tended to his heavy correspondence, and his students both in and out of the classroom. The indefatigable Smith was now entering his sixty-second year, yet these accomplishments merely hint at the scope of his activities.

4

The Fourth Decade, 1904–13

1904

Bibb, Bullock, Coffee, Covington, Crenshaw, Dale, Escambia, Geneva Greene, Montgomery, Pike, Sumter, and Tuscaloosa Counties

The 1904 summer trip got underway about the usual time with Truman as driver and Old John and a new mule named Joe pulling the wagon. Merrill came along as odd-jobs boy. As usual, last minute trouble developed—two essential bolts pulled apart in the back spring, and they had to take the wagon into Tuscaloosa for repairs. They then headed for the Hardy Clements place where they planned to camp, but after passing through Cottondale they "found the roads very heavy from recent rains and [Joe] began to show signs of wanting to stop and meditate whenever we struck a particularly tough piece of road and a steep hill. Within three or four miles of Clements to add to the trouble they had just worked the road and the pull up steep long hills over loose dirt or half mud ten inches deep was something terrific. Before reaching camp I had about decided that we would have to send [Joe] back.

"During the night, about ten o'clock there came up a great show of a storm, with incessant lightening and some thunder gradually approaching nearer but it did not rain very much, only enough to wet up everything and keep us all awake for an hour or two . . . This morning Mr. Fleet Clements who lives at the old place came with us a mile to show us one of Mr. Pinckard's ore pits . . . It is interesting to me to see the ore this far SW.

"Mr. Ward drove up the big hill and gave Truman a lesson in driving. [Joe] stopped only once, and has done well all day, a good puller, but he has evidently been over loaded. I think we will make out all right with him.

"About four miles from camp we were misdirected and spent about four hours out on a road leading west to the RR. In getting up a sandy hill where a deep hole had been washed we came to a standstill, and broke a trace in the bargain. This we spliced with staples and came on back to the main road."

Near Centreville they camped in the woods among two sawmills and the M & O Railroad, with clearing skies overhead and the muddy roads left behind ". . . for which we cannot be too thankful, especially the mules," Smith wrote in his journal.

On July 14, they rolled into Prattville about seven p.m. after getting lost twice due to poor directions. Smith had found, when asking for directions, that people usually would say, ". . . 'keep the straight forward road.' Not one in a hundred will ever tell you which fork to take when the road divides." They spent four days in Prattville allowing time for Truman and Merrill to visit with the relatives and Smith to go to Montgomery to push for the printing of his report on cement resources. He also bought supplies for the wagon, and upon his return the expedition left for Mt. Meigs about four p.m. "on a first class road."

When they arrived in Mt. Meigs three hours later, clouds were roiling in, rain was imminent, " . . . and to cap it all we could find no one to take us in and no place for camping. At last just at dark and when the rain began to fall Mr. Bonnet asked us to drive into his lot and camp which we did. His house was full so that he could not take us in, but the rain soon ceased and we got on all right, Truman sleeping in the barn so that we had no need to stretch the tent."

They camped two miles south of Aberfoil on July 20. ". . . We came up to Union Springs this a.m. by ten o'clock and met a lot of University folks, and spent the morning answering letters that have come to this place . . . We took dinner with the Normans and met Eley, Rainer, Gaines, Pope, Fort, Davis, Frazier of war times, Mosely, etc."

The next day on the way to Troy, ". . . we ploughed through the deep sands of Aberfoil . . . From Perote the road, contrary to my expectations was a cross country one up one hill and down another with no flat lands between, or if any they were crossed with sand. On most of the hill tops 18–14 miles of Troy iron crusts and small pebbles of limonite cover the ground and these broken up crusts down sandy slopes and up them make traveling very disagreeable. After passing Catalpa about twelve miles from Troy and for that matter for several miles before getting to Catalpa we have beds of deep white sand equal to Ozark and after passing Catalpa we get upon very high ground and everywhere a bed of white or gray sand about five feet deep over the red sandy clay or loam. For seven or eight miles, perhaps more, we have these deep sands up hill and down and heavier pulling I have never seen mules do . . . I am at a loss to know what all this sand capping is . . . The prevailing timber here as yesterday is composed of the three black jacks."

On July 22 Smith was in Troy: "[At Glenwood, Crenshaw County,] I got a team and drove out to Mr. Abner M. Jackson's on Deer Creek, a small tributary of Conecuh River. Near his house is a very good exposure of the Coal Bluff, Landrum's Creek lignite of which about three feet show above the bed of the branch or creek in one place. The bed of the creek there is lignite and I don't know how thick it is. The young men who went with me, sons of Mr. Jackson, the old man being off fishing, say that in a well bored at the house they went through thirty feet of lignite and got water in the sands below. I noticed, however, that they made no distinction between the black clay of which two to three feet thickness overlay the lignite and the lignite itself, and I imagine a good part of the thirty feet was black clay."

On July 24, they were camped at Flemings Mill at the bridge over Pea River. It was a Sunday, and that morning "after breakfast we walked out to the depot where a great well 24 feet square and sixty or more feet deep had been dug by the RR, in vain for no water was obtained. The material thrown out of the well, however, supplied many good fossils. . . . I got a cigar box full which I expressed to Mr. Aldrich.

"We went to the Methodist Church where we heard a sermon from Mr. Peavy with whom we afterwards dined."

They reached Ozark by July 25: "We got to town about one, not stopping for lunch and were fortunate in getting a camp in a vacant lot with shade trees one block east of the Courthouse Square. . . . Mr. Reynolds drove me down to the Whaly Bridge over Choctawhatchee seven and one half miles from here to see a few small concretions of pyrite in the laminated clayey sands that make up the bed of the river there . . . [He] . . . promises to have a number of these taken up and shipped to Tuscaloosa.

"We got back about dark and found that the boys [Truman and Merrill] had prepared supper and were waiting."

July 26: "We came on a mile this side of Haw Ridge to Mr. Harding's where we camp at the turning off of the Enterprise road. I have enquired about members of the 33rd Alabama Regiment, but find very few surviving around here. I met a son of Capt. Hughes at Haw Ridge. Dr. Lee's home was pointed out."

The following day they were in camp four miles east of Elba. ". . . We came through Clintonville and while there a man passed in his buggy in which was a large sack with the ice from the hail storm ten days ago. It was in the form of conglomerated round hail stones some half to three fourths inches in diameter."

July 28 at Coffee Springs: "About three miles from Brockton we came into the worst territory of the hailstorm of the 18th, and a sight it presented. The cotton is completely stripped of leaves and bolls, and nothing left but the bare brown stalk. Where the destruction is less complete a few leaves have put out since the storm but there are many fields in which the cotton looks exactly as it does in mid winter. The corn also at its worst is stripped of leaves, many of the stalks beaten down and ears beaten off. In more favorable places the leaves are torn to shreds but still hang to the stalk. Sugar cane and sorghum and millet are less injured but still torn to shreds. The trees of the forest also present the appearance of midwinter for leaves are mostly stripped off, and the remaining ones are brown and withered. Seen from neighboring hills these woods where most injured show plainly by the brown color and bare limbs. At the intersection of a Clintonville-Geneva road with ours there seemed to be most destruction and into a ravine nearby a great flood of water from the heavy rains that accompanied the storm washed tons of hailstones, here they remained till last night 27th, nine days from the date of the storm. I saw yesterday a bushel or so in a buggy brought from the place, but this morning I could find none, and only a few pools of ice cold water where they had been. The rain of last night finished them. The houses in this district show a peculiar appearance for they are completely spotted on the exposed sides, the hailstones knocking away portions of the old weathered surface exposing new wood. So also on exposed sides few window panes remain. As we come out to Enterprise we leave this region of destruction."

The red sands of Enterprise interested Smith: "These sands . . . cement together in a remarkable way and when thrown upon the street soon become almost stone hard. I cannot imagine anything better for road material especially if used in connection with gravel. [Mr. Rawls] showed me some good looking fire brick made from clay of a shallow pond near town and will send samples for testing . . . and for the Museum also in Sept.

". . . The roads are good and hard and with many level spaces and as a matter of course the turpentiners and saw mill men have wrought destruction to the forest. A train road from Richburg is carrying out the pines by the score every day. We get into Coffee Springs about six-thirty and camp in a beautiful grove in the upper part of the place near the Spring and Pool."

Before leaving next morning, Smith "went down to the Spring where they have made a swimming pool. At the bottom of this pool there are four or five springs boiling up through the sand, the boils are two to three feet across. The drinking water is from another spring over which a large tile drain pipe has been placed."

The wagon made its way from Coffee Springs to Geneva by noon. They passed "through tolerably thickly settled region, the houses all very good and the people apparently prosperous." They camped in a grove on "a beautiful level spot . . . good for camp barring flies and gnats and an occasional calf."

Florala won praises from Smith: "Florala is a flourishing place and most beautifully situated on McDavid's Pond. As we drove up the wide avenue which has been laid out along the route we came, we crossed the ditch which in 1891 someone was just cutting for the purpose of turning the waters of the pond into a stream to the north, with a view to using the fall to turn a wheel. I am told the project failed and the tree grown ditch seems to be evidence of it. I came out along the borders of the Lake past the place where I took the photograph in 1891, but the surroundings are very different, for then there were no houses here except one or perhaps two stores."

On August 1 at Camp Allen Hart five miles northwest of Oak Grove, Florida: "Our John mule was under the weather today and would not eat at all, which has put us all out of our reckoning. We came on to Laurel Hill which seems to be a flourishing RR town thence to Oak Grove where we crossed the river on a new bridge. Just below the bridge on the right bank is the log landing where the bed of Miocene shells is exposed. This place shows about ten to twelve ft. the lower half of which is a greenish or grayish sand packed with shells in beautiful state of preservation and not soft and breakable as is so often the case. The upper half of the bank is of river sand. Merrill and I stopped here while Truman drove on up to Mr. Franklin King's who is post master of Oak Grove office, and unhitched for noon. We collected a fine lot of shells . . . All the road traveled today has lacked the open pine barren appearance, and on the other hand red clay or loam hills of low elevation are characteristic. Over the red loam is usually a gray sand supporting the black jack growth."

John the mule continued to refuse food, "and we were at a loss as to our best course, but finally decided to drive on. A rain came up which complicated matters. After leaving Oak Grove we travel up the river along a public road for a couple of miles, then turn off westward by private road, which we keep for ten miles and also because of the fact that for the next six miles or so the road goes through the pine barrens where there is not a house and we were afraid to risk it with a sick mule." With what must have been a deep sigh of relief, Smith added, "To our great satisfaction John has taken to eating oats."

In Brewton on August 3, they "got a fine camp in the yard of the old court house." Smith spent his time there answering the twenty to thirty letters waiting for him, visiting with many friends, and making plans. August 5 was another day of mishaps. They were advised "to take left hand up river road, but when we came to Dixons Mill Creek we found it impassible by reason of logs washed in and quicksands and deep holes. We were therefore compelled to turn back and we had a time turning the wagon as there was only the narrow road and deep sand. One breast strap broke in trying to turn around and we finally had to do the turning by hand and it was a job sure enough. Meanwhile, of course, it clouded up and rain threatened on all sides. Came back to the regular Douglassville road after losing an hour or two and came out it to Mr. Dixons . . . Here we fed the mules and went on down to Coal Bluff, a walk of about a mile."

The next day at Flomaton, Smith sought out the only artesian well in town. It was 311 feet deep, Mr. Ivey told him, and the water rose eight to ten feet above the ground. When at first a smaller pipe had been used, the water cleared the top of the full-grown sycamore tree in Ivey's yard. The water tasted mildly sulphuric with a dash of iron. White sand was the predominant material removed, though toward the bottom Ivey said they brought up a narrow, three inch long oak leaf, petrified, and some charcoal. "He also tells me that at Old Erie (between Flomaton and Pollard) they have or had a flowing well, and that they struck a 'cypress' log at 80 feet depth below the surface."

The geology wagon rolled into Pensacola on August 8 over a "first rate road in which there were only one or two turns and for four or five miles the road bed was made of the waste from burning of pyrite and was hard and firm. At 1:50 we went with the dummy down to Fort Barrancas and on returning spent some time on the wharves looking at the shipping."

The August 17 Daphne newspaper carried an account of the Smiths' visit there. They had set up their camp "on the bay front . . . on Thursday of last week and left on the *Heroine* on the following morning for Mobile. Had a two mule team and all the camp outfit needed to make them comfortable . . . The doctor has been looking up the waters of this section, especially natural springs and other drinking waters. The editor had [the] pleasure of renewing a friendship formed over 20 yrs ago in Tuscaloosa."

From south Alabama, Smith angled toward Mississippi. They camped at a turpentine still near Leakesville. "We found that along this road no one lives for the next ten miles so we concluded to go into camp here at four and fortunately, for Old John Mule has the colic again." On August 20, they were in camp just north of Waynesboro, Mississippi. "We had some trouble and delay at a little creek three to four miles south of the bridge over Big Creek for we drove down the bank—steep—and into the creek before we found out that it was deep enough to come up into the wagon. We therefore took the mules out and undressed more or less and backed out up the bank by hand after one and a half hours hard work for the clays forming the bank gave no foot hold and we had to dig out each footstep as we progressed."

On Sunday the twenty-first, they traveled by way of Shubuta to overnight near Pachuta. "At Mr. Dabb's we were very pleasantly situated, with a good grove to camp in and plenty of oats and hay for John and Joe." On the next day, they stopped "till after dinner with Maj. Berry where Judge Johnson is stopping. We then came on to Enterprise [Mississippi], the Judge and myself going to Watubbee station to see the cuts showing the gray clays above the Claiborne.

"At Enterprise we found all the hotels filled to overflowing so that we could not get our supper and breakfast there and we therefore came up to the old camp where we fared very well . . . Camped in a branch bottom, where a little rain will flood us. No sign of rain yet but we confidently expect it before morning." On August 24, they were back in Alabama at York where they "stopped at the Moore House overnight . . . I lost my glasses very near the place where I lost a similar pair two years ago.

". . . Dreading the crossing at Jones' Bluff, I decided to come around by the Gainesville ferry especially when told that Gainesville was only 25 miles from York which is true enough but no allowance was made for the fact that the first ten or twelve miles of the road have a NW and W direction so that at Payneville we were only about two and a half miles of the Miss. Line and nearly as far from Gainesville as York is."

They had to make camp near Sumterville for the night of August 25. "Sumterville is perched about on a number of hills . . . The stores are on a narrow ridge. Coming up from the south we seem to be coming to the jumping off place, and so it almost is for the descent begins at once when we reach the summit . . . Mr. Joseph H. Wrenn of Sumterville has a house and plantation four and a fourth miles from town and he has kindly given us leave to camp here and to get feed etc from his factotum Ernest. We therefore kept on till after dark but have fared very well notwithstanding. Rain has been falling this afternoon in every direction and black clouds have been overhead, but since we have come into camp, things look better. We have had the worst of roads today. Through flatwoods of course. The clay is much cut up and rutty, and the same is true of the prairies, so that we have been on a continuous jolt much of the time . . ."

Safely back in Tuscaloosa on August 27, Smith summarized the harrowing last day of their tour: "Yesterday we drove up to Gainesville and out to Mt. Hebron by 11 a.m. There we stopped for dinner for several hours with Capt. Poyner [Smith's friend for over forty years]. After dinner we came on as far as Union, passing through Clinton with its many artesian wells. The town has recently (April) been practically destroyed as almost all the business houses have been burned to the ground. Halfway between Clinton and Union there came up a sudden storm of wind, rain and hail which caused our mules to turn aside with a view to presenting their hinder parts to the storm. As this would have resulted in the breaking of the pole or harness or both, Truman had to jump out and unhitch and stand in the rain and hail till it was over. Of course, he was drenched in half a minute. As the wind came from left front of the wagon it blew the rain in, breaking loose the front curtains and I also was soon saturated in front. Merrill in a great panic, jumped out of the wagon and caught hold of the hind wheel and began crying out for dear life.

"Hitching up and driving on after the storm we found directly three large pine trees across the road and had to turn out through a field of tall cotton with water in all the ruts. The roads were very heavy as a matter of course, and we had difficulty in getting up some of them and also in getting up on a bridge with steep slope as the mules could get no foothold. All this delayed us and it was after seven when we got to Union and got shelter with Mr. Crosby."

Even in the midst of pandemonium, Smith's alert geologist's eyes "noticed a fine silicified log, some six or eight feet long and two feet through embedded in the sandy clays by the road side in a cut going up one of the numerous steep clay hills. It was nearly dark but I thought I could make out that some of the material was agatized." This arduous trip may have had a bearing on the retirement of the geology wagon at the close of the 1904 season.

Smith's long-time associate and friend, Henry McCalley, died unexpectedly November 21, 1904. Returning from the St. Louis World's Fair, McCalley stopped off in Huntsville to visit his sister. He fell ill with what doctors first thought was malaria, but he died very quickly from pneumonia. McCalley had come to work for the Geological Survey as a volunteer in the summer of 1877. The next year, he became an assistant in the chemistry department. In 1883 he was appointed assistant state geologist. He served Alabama and the university almost thirty years, according to a newspaper of the time, and "was almost as well known over the state as E. A. Smith to whom he was so able an assistant—a quiet man, very forceful, decided character, a fine business judgement, owned considerable property, . . . a member of the Episcopal Church, a vestryman, . . . graduate of University of Virginia, a Master of Arts from University of Alabama."

McCalley wrote or contributed to a number of important reports for the Geological Survey and many honors came to him during his lifetime. One newspaper account added a sad footnote to McCalley's professional life: "[He] had been engaged during the last two years of his life in the study of the gold regions. He was very nearly ready to put his results on the map and to prepare his report when he died. It is to be feared that a good part of his field notes will be difficult or impossible to interpret and that much of the work will have to be done over."

Smith closed his own tribute to McCalley in this way: "It is said of him that as a child and a young boy he had never received a correction from either parent or teacher and he never neglected a task set before him; these characteristics followed him through life."

1905

CHOCTAW, CRENSHAW, DALLAS, AND MONROE COUNTIES

Landon and Truman were both working away from home but in different towns. Julia, who lived with their parents and taught languages at Tuscaloosa High School, often wrote to them. Her handwriting, legible and without frills, made the words march straight across the page as she cheered her brothers on in an affectionate, funning way. "The boys," especially Truman, needed a cheerleader, for they grumbled about their work and changed jobs often. They were always short of money and usually having romantic difficulties.

Julia's pet names for Truman, nine years her junior, included kid, kidlet, dear old Truman. She wrote him in Berkson, Mississippi, June 11, 1905:

> Dear old Truman.
>
> I have thought I would write to you every day since you left. But after the commencement rush Emma [the Smith's cook] was sick for a week or more and I spent most of my time fooling around in the kitchen. Then Sue and Gussie stayed with me a few days, and yesterday and today Nan Lewis [Julia's cousin] has been here. Papa gave her your [mathematical] problem and she has been working all the morning on it and is still at it, but she don't want Prof. Palmer [professor of mathematics and long time friend of the family] to see it till she has either gotten it or given up.
>
> . . . Emma has at last gone back to cooking, and we feel like ladies of leisure once more. We all miss you very much indeed. Little old Punkie [Merrill] is still in Prattville and I don't know when he is coming home. He wrote he had eaten a lot of plums but they did not hurt him so Mamma need not be afraid, and that he had missed brushing his teeth two nights but that he did not think it hurt them much.
>
> I wish Papa was not working so hard on his old report. He seems so harassed and nervous that he is hardly like his old self. He eats almost

nothing etc. But you know how it has been with him for a month or so. I hope he can take it a little easier this summer. I hope you will have a good time all summer and will get down to Mobile once or twice [to see his sweetheart]. Write when you can.

With much love from us all,
Devotedly,
Julia

Within a week of Julia's letter, her father was in Luverne visiting ". . . a place worth coming to see"—Brunson's Limesink three miles out of town. This "great sink," at least thirty feet deep with a diameter of perhaps two hundred feet, plainly displayed in its sides three sections that Smith described in his journal. From the top of the sink ". . . a small branch descends by one or two falls over the harder beds . . . to the bottom . . . where it disappears through a crack or cavern—obscured . . . by the sands."

He also visited a quarry and lime kiln at Luverne, guided by "a young boy Sanders, son of one of our University students of 1881 . . . From this quarry has been obtained chimney rock and rock for lime burning. The material on casual glance can't be told from the St. Stephens chimney rock, except that no orbitiodes show in it. It is quarried with saw precisely like the St. Stephens, dressed into blocks for building, the same shows up in chimneys, and some walls of houses in Luverne, just like the St. Stephens and is burned into lime in kilns constructed by the same material. I spent an hour or two breaking open the old blocks hoping each moment to disclose a well-defined and recognizable shell, but to no purpose though the material is full of fossils and fragments. Perhaps an oyster or an Umbulla might be identified. I shall turn the matter over to Mr. Aldrich."

In mid-July, Smith left Tuscaloosa by train at 7:30 a.m., reached Selma by noon where he spent the night. On to Tunnel Springs next day to visit Dr. Nettles, only to find he lived one station farther on. Once arrived, he discovered that Dr. Nettles and nearly all the men of the surrounding neighborhood were in Monroeville to appear in court as witnesses. "So I had nothing to do but wait over till night, when the Dr. returned but had to go after supper six miles out to a patient."

On July 20, "Dr. Nettles had calls on him which prevented him from coming along, but he kindly placed his two horse outfit and driver at my disposal and we have rolled over the country at a great rate at least five miles an hour . . . [At Kempsville P.O] . . . all the men are in Monroeville. Coming back by the station we came out to Dr. Nettles' father's store and mill and house where we stop. The Dr. cannot join us tonight nor tomorrow because of the court and his father also is away at some place. There are ten of the Nettles boys here, counting the Dr., and five girls, several married and living nearby."

At the Bladon Springs Hotel on July 25, Smith "went direct to Cullom and took a bath in the salt water of the bored well and returned to Bladon. Cullom is not now being kept up though Mr. Barnes takes in those coming here into his house just outside the hotel yard. It is rather depressing to see the big hotel unoccupied and going down." The next day Smith spent "the day about the hotel there being no chance to get buggy or conveyance or guide. I get some interesting data about the salt licks and mineral springs about here.

"Salt Pond is such a suck or lick an acre or two in extent, a bare spot no vegetation, a crust forms which will be for a time firm enough to walk over, but directly bulges up in the middle and cracks open and out of cracks oozes a black mud. One can thrust a stick down it, consisting of mush. The crust can be shaken by standing on it and jumping up and down. Salt wells bored long ago now deliver into the flat. The Salt Pond thus has its name formerly a beaver pond (now drained) in which the stream from salt wells flowed.

". . . Turkey Creek seems to be an interesting stream for both branches of it have these 'Sucks' as they are called much sought after by cattle which are fenced out where possible because of danger of miring up when crusts break. Dansby's Springs . . . is a collection of iron, sulphur, and soda springs. The crusts in these sucks are covered in dry weather by an incrustation of Carb. of Soda according to popular belief and Mr. Turner says his father's cook has made bread with it as a substitute for the usual Sod. Carb . . . . Back of Conner's store there used to be an alum spring now covered up . . . A well nearby was so strong of alum that it was disused and filled up."

While Smith was with the Nettles family and at Bladon Springs, Jennie, Julia, and Merrill stayed close to home. Julia's plan to visit a friend in Mobile had to be cancelled because the city was under yellow fever quarantine. Much to Jennie's distress, the dummy, running from Tuscaloosa to Holt past the Smith house, ran through her turkey flock, killing two and hurting two so seriously one died and the other had to be killed. In August, however, matters closer to her heart upset Jennie. Truman, in love with a university coed named Pearl Boyles from Mobile, badgered his mother to take Pearl in as a boarder for the new school term. Jennie explained tactfully in several letters why that would be impossible—there was no place to put her except with Julia, and Julia did not want anyone else in her room, and besides, her room was too small and crowded for two occupants.

On September 14, the *Montgomery Advertiser* carried an interview with Smith in which he showed the reporter through rooms stacked with the university's collections. From a locked case, Dr. Smith took "half a dozen lovely stones of a robin's egg blue color. 'These,'" he said, "'are turquoises. You will observe that they have taken on a beautiful polish. They were found by me in Clay County on a farm at Franklin . . . I am absolutely convinced that there is good money to be made in the gathering and selling of these Clay County turquoises. The owners of the land on which they are found do not seem inclined to take advantage of the opportunities that have been pointed out to them.'"

In December, Smith wrote Truman in Macon, Mississippi, of his disappointment with Truman's decision to spend Christmas in Mobile with Pearl and her family, but he realized Truman would enjoy that more than coming home. On the last day of December, Julia wrote a fond letter to Truman in Mobile wishing that he and Landon had come home for New Year. She told him that Mama was with Papa in New Orleans, and that Papa wrote that Mama was having such a good time she did not want to come home at all. Julia assured Truman that she and Merrill were having a fine time also—they were invited out so often they had given Emma, the cook, a holiday, and Julia had friends in for chocolates and salad. What Julia did not mention to Truman was that she had given her mother thirty dollars to make the trip with their father.

1906

AUTAUGA, GENEVA, HOUSTON, JEFFERSON, MADISON, TALLADEGA, AND TUSCALOOSA COUNTIES

Landon was now working for the M & O Railroad in Waynesboro, Mississippi. In February, Truman, employed in Holt, requested that he get passes on his railroad for Pearl and himself. Landon demurred, however, as he was only a temporary employee. He advised Truman to go to work for the railroad again, so he could apply for his own passes.

Smith went with Merrill on his March class trip beginning in Birmingham "out 20th St. to foot of Red Mtn and walked thence up to the top and over to the Sub. Carb. Chert quarries on eastern side. Then down the mtn seeing the iron ore mines of the Central Coal and Iron Co. together with some old workings, now not used. From the Red Mtn we walked back through Glen Iris neighborhood . . . taking the street car . . . for town. After lunch we went out to Ensley where under the guidance of Mr. J. R. Harris we went through the steel works.

"Tuesday morning we went by Central of Georgia train out to Sylacauga where we took a hack and drove down to the old Gant quarry now being extensively worked by the Ala. White Marble Co. I was not expecting to see so much work done as they have accomplished. They have a mill for sawing and smoothing the blocks and slabs etc. together with polishers. They are getting out many very large blocks but there are streaks of slate . . . running through some of them but it is possible by selecting to get some fine pieces free of flaws.

". . . We took the train for Alexander City where we arrived at 7 p.m. On Wed. we got a surry for the gold mines going by Dutch Bend where we inspected the cyanide plant at the old Ulrich place. A stamp mill beats up the ore in a cyanide solution and the slime or pulp is then conveyed to cyanide tanks for better solution and the gold solution then carried to the vats to be precipitated by zinc shavings. [In Alexander City that night] we forded the flooded streets in a pouring down rain and took [the] train back to Birmingham."

Landon sent Truman the passes after all and enclosed a twenty-five dollar check "as per your request." On June 10, Smith wrote Truman about a possible job for him with Mr. Buck in Mobile. He added a complaint against Julia:

"Julia expects to try that examination tomorrow but from what Mr. Foster has told her it may last several days. In that case I have advised her to drop it all for she cannot afford to spend every day till she leaves for Europe in getting ready for that exam. She will have to take a little while to get ready for the trip. Besides the exams are mostly on subjects which have no connection with French and German, and these two occupy a very small fraction of the time set for the examination, so I think it is a pity that she has wasted so much time on it all." Later Smith reported to Truman: "Julia was in examination yesterday till after 8 p.m. and today will be there till noon. I wish she had let the thing alone."

That same day, Smith wrote Landon, still with the railroad but now in Jackson, Tennessee, asking for more passes for Truman and Pearl. Landon protested that he had just asked for passes for Mamma and Merrill to Montgomery and "don't wish to ask for another one so soon." When Landon heard that Truman got the Mobile job with Mr. Buck, he warned Truman that his own boss, Mr. Buck's brother, said the Mobile Buck "liked to get someone else to do his work for him." Landon added, "But I feel sure you will get along with him okay and it will be good for you if he lets you do most of the work."

On June 22, Jennie wrote Truman: "I am sorry that you would have felt worried or troubled by anything that passed between us. There is no question of forgiveness, there is nothing to forgive, we were both a little worried that is all. I believe after all if it is not too late the best thing for you to do is to try Buck. That is[,] not to accept the position with the Southern [Railroad]." She went on to tell him that Cousin Rosebud Lewis's husband said working for the Southern "will injure any engineer's reputation . . . , they do nothing but the cheapest and shoddiest kind of work." [Smith, however, discounted this story in a letter to Truman.] "Your Papa got off on the 7 o'clock train yesterday morning," Jennie continued, "so I am now all by myself. I've not had to eat a meal alone yet though for I had the Lewis crowd out to spend yesterday with me, and Mr. Stickney came over in time to take tea with me . . . I have so much to do and keep so busy I do not feel lonely but at mealtimes for I do hate to eat alone, and then at dark it begins to feel very lonesome but I shall manage somehow. . . . With dear love your devoted mother."

Smith's train trip Jennie mentioned was to western Mississippi with several others for collecting shells. He wrote in his journal, "Mr. Aldrich has been prevented from coming and thus we miss the very great assistance which his knowledge of shells would give us."

Julia had departed for Europe on the German ocean liner *Bremen* with Eugene's sister Julia Pratt. "Julia wrote us a short note from on board the ship—," Jennie wrote to Truman on June 26, saying that "she got about two dozen steamer letters and a box of flowers and the finest bunch of American Beauty roses that were on board the ship. Both Mr. Yuille and Dr. Hope had given her boxes of candie which Merrill says he knows she will have no use for on the ship. And I think the same."

Smith wrote Truman that he could no longer afford to keep the houseboat, but the decision to sell it made him feel "half homesick." He asked Truman to help him identify a photograph of an old cotton press. He may have taken it "the day we left Dr. Moore's Hotel at York on our way up towards Sumterville, but I cannot find any mention of it in my notes of 1904, which was our last trip together. Mr. Hodges says it was not taken on the trip when he and I and Merrill made the party, and I hardly think it was as far back as 1900 when you and Hodges were along the year you had fever down at Thomaston. If you can recall the picture try to place it for me." (After the 1904 trip, the geology wagon had been retired, and Smith used the train and the automobile.) He added that he had had Truman's canoe brought to the house for storage and wrote in minute detail a description of the canoe's condition and a new plan he had for hanging the boat.

Soon afterward, Smith left for southwest Georgia on a collecting trip with Aldrich and a geologist from the Georgia Geological Survey. Meanwhile Jennie, in Prattville, wrote Truman a reminder to tell young Julia about the "nice letter" her friend, Mr. Yuille, wrote to Mr. Buck on Truman's behalf when he was applying for his present job.

In mid-July, Smith warned the ardent Truman: "I'm afraid you do not give [Pearl's family] . . . any rest. I would not crowd them too much if I were you. You might spoil your chances." As the fall school term approached, he confided to Truman: ". . . I wish you were going to be back another year. It will seem strange to me not to have a boy in college as I have had for nearly ten years, in fact it makes me feel lonesome and almost out of place." Smith, unflagging supporter of the school's sports program, must have been proud of the athletic record of his two older sons. Landon had trained to be an excellent swimmer even from the days of his summer travels with his father. Truman starred on the university's baseball and football teams. A Birmingham newspaper called him "one of the best all-round athletes in Alabama's history." Smith's support of every sport played at the university was unflagging, although baseball was his favorite. He seldom missed a practice session and attended the home games when possible.

Pearl was returning to school to earn an advanced degree. From Prattville Jennie reprimanded Truman: "I am sorry you have planned your visit home at the opening of the University as I will not be at home at that time. I will have to stay with Dora till her mother returns [from Europe] which will be after the middle of September. Your Papa of course will be at home and Emma has promised to be there to cook and take care of him for me until I get back—but you certainly could not entertain Pearl Boyles if that was your plan. With dear love your mother."

Before the fall term began, Smith was in New Orleans traveling along the Gulf coast by "the electric line, a most pleasant route running between the houses and the water's edge. Passed in front of Beauvoir [former home of Jefferson Davis] now the Miss. Old soldiers' home and on to Gulfport, a new town and of no interest . . . Spent the day in Mobile till Sun. night meeting Landon and Truman . . ." Back in Tuscaloosa, he wrote Truman urging him to take an electrical engineering course at Perdue; "Papa" thought that would make Truman happier in his profession and it would be a step upward.

Smith wrote to Truman on August 11: "I am forwarding this morning a . . . postal . . . from Julia. Her address is the same as heretofore, although the [tour] . . . ended in Paris yesterday. They start on a new trip with the Archibalds after spending a week in Paris, and will go to Belgium and Holland and afterwards to England and Scotland, setting sail from Liverpool on the 12th of September. I think your Aunt Julia wants to spend a little time after returning in visiting some of the northern kinfolks and in that case if Julia stays with her they will not be home before the first of October. I hope Julia will stay, for it can be arranged about her class in the Public School."

Julia's letter to Truman, which Smith enclosed, had been written aboard ship on June 27, 1906: "I certainly wish you were on board, for I believe you would enjoy a sea voyage on a large vessel like this more than any one I know. Eight times around one deck is one mile and this deck is not much more than a third the length of the whole vessel. I know too how you would delight in the machinery and everything concerned with the running of such a big boat, to say nothing of the sunsets, moonlight on the water etc, etc. We have had [a] . . . very smooth passage and I have not been very seasick but I will be very glad when we have something steady under us once more. Our party seem to be pleasant enough but not specially exciting. The young man is pretty boring and pretty conceited. How are you and Pearl getting on? I hope you got a job some where near Mobile and are happy once more."

On August 20, Smith was in Dothan marveling over the growth taking place in southeast Alabama. "It is a matter of some surprise how these . . . towns have sprung up. Hartford and Slocomb are both promising looking places while Dothan is a large and very rapidly growing town. When I drove through this section in 1880 on the way to Fla. there was no suspicion of a Dothan. In 1891 it amounted to very little. Now it has a population of 8000 to 10,000 . . ."

He had a similar reaction when he arrived in Huntsville: "Mr. Pettus drove me around all morning to give an idea of the growth of Huntsville. We went first westward about three miles by several large cotton mills to the place of Judge Stewart where there is a bored well which yields a villainous water containing over 600 grains per gal. of solid matters . . . The solids are mainly Sodium sulphate, sod. Chloride, Calc. Sulphate, Magn. Sulphate (in the order of their abundance.) This should be a medicinal water."

Smith wrote Truman a lengthy letter on September 30 dealing with Truman's affairs, the storm that had just hit Mobile, the school coach down with malaria, and other subjects, but about Julia's return from Europe he had few words: "Julia got home yesterday afternoon by the M & O and we shall be glad to see you up here whenever you can get away . . ."

In mid-October, Smith focused his attention more intensely on his two older sons; he looked forward to a visit home from Landon. Smith arranged to see a man he knew regarding a job for Truman on the Southern Railroad, he found another possible job for him on the Missouri Pacific, and he worried that Truman would quit his present job prematurely.

On December 16, in reply to two letters from Truman, Smith wrote: "I hardly know how to advise you in the matter which you ask me about, for I have already given you my views on the subject about as fully as I could. You know you are more to me than almost any one in the world, and that there is nothing I would not do to promote your happiness. You know also that neither your mama or myself has any objection to Pearl, how could we have? In all our advice to you not to marry too early, we have been influenced by what seems to us the interest of both of you. You remember I was in love with your mama for some 8 or 9 years before we were married, so that a few years waiting is not so very serious a matter, if both of you remain constant. Then again teaching does not harm any one, but, the contrary, a short discipline of that kind might often prove beneficial to any one, as it would cause him or her to meet and adapt himself or herself to new responsibilities.

"I cannot help always thinking of the case of Mr. Anderson here [an assistant in the museum], who began his married life on some $60 a month. With the necessity of providing for a family even though a small one, he has been prevented from getting into much better paying position, and the consequence has been that he has had a continual scuffle to live, and the prospect of any improvement seems to me very remote . . . [I]f you were in a good well established position with prospect of advancement, or at any rate, with no probability of your being thrown out of a job or reduced in pay by a change of officials, say, then there would be no particular reason for not marrying. I would be certain, however, not to have any unpaid debt behind me, and I would not like to think of starting out without a reasonable sum in advance. For expenses will come up that you never dreamed of and which no one could foresee. I would also strongly advise every one to have a small amount at least of life insurance to his credit before he married. I did not myself follow the advice which I am giving you for I did not take out life insurance till after I was married, but I had a good position, and could offer your mama a good home here at the University with reasonable prospect of permanence.

"After a man has the responsibility of a family he cannot have the opportunity of preparing himself any further to make the most of his profession, for he will be obliged to take whatever position is offered and will have no choice and no chance to put off the undesirable for something better.

"If I could have things as I would like, I should have you fully posted and educated in every branch of engineering, or at least I would give you the opportunity of getting some idea of all branches of the profession, and of making an intelligent selection of that for which you had most taste and inclination and

(Opposite page) Geological survey camp near Helena, Shelby County, 1907.

adaptation. As it is you have had an insight into a very small part only of the engineering, and might with the opportunity given, discover that there were other parts of the profession which would suit you far better, and in which you could rise to some eminence. I hate to think of your being always tied down to some restricted part of engineering without the reasonable chance of becoming independent.

". . . But, why tell you what I have so often told you? I hope to see you before very long and then we can talk matters over very much better than they can be discussed by letter. I may leave here before Christmas or on Christmas day, so the prospect is that you will not get here till I am away."

Unmentioned in this letter was the fact that Smith, as of 1906, was in a stronger financial position than he had ever been. For the first time the Alabama legislature earmarked fifteen hundred dollars for the state geologist's annual salary. In fairness, Smith's lack of a salary since 1873 was not all the legislature's fault. Smith himself felt the money could be spent in other ways more beneficial to the state. Now, however, the pressing financial demands of his family may have changed his mind.

On December 24 in a letter to Truman, Smith commented: "The buggy, about which I wrote, has come in and we have set it up and made one trip in it, at least Julia has. It is a nice present and it is made by your Aunt Julia and Cousin Dora to your mama and Julia. I am afraid that your mama will not get much benefit out of it, because she is so afraid of the horse. Julia, however, will get the good of it. It has rubber tires and is altogether a pretty turn-out. It is pretty high from the ground which makes it hard for your mama to get in and out." Smith himself was ready to leave that day for New York and the eighteenth annual meeting of the Geological Society of America in Ottawa, Canada, where he would be elected vice president.

1907

CLARKE, COFFEE, COOSA, CRENSHAW, GENEVA, HOUSTON, JEFFERSON, MADISON, MOBILE, MONTGOMERY, TALLADEGA, AND TUSCALOOSA COUNTIES

Plans were well underway by 1907 for the building of the "Greater University." The first one hundred thousand dollars of the legislative appropriation was earmarked for a biology-geology building housing,

As the survey crew traveled in summer, they often happened on baptisms. This one was at Ragland, St. Clair County, 1907.

among other things, laboratories, an office for Smith, and his geological collections. Though the cornerstone of what was to become Smith Hall was laid in June 1907, the building contractor was not hired until March 1908, and the building was not ready for use until the fall of 1909.

Despite Smith's many interests and endeavors, he continued to worry about Truman and Pearl. On April 10, he wrote Truman, now working in Hattiesburg, Mississippi: "As to Pearl's teaching a year I cannot see that it would spoil her at all, it might be a very good thing, and so long as you are not in position to get married and set up in housekeeping I don't see why such strong objections on your part."

Smith next turned up in Elba, looking at overflowing wells and lime sinks and visiting old friends. He admired the "good graded roads" being built in Pike County. One morning he visited a group of young people from Troy (perhaps from the Baptist Children's Home) at the "Red Bug" camp on the Conecuh River. He made photographs of the camp and of the boys and girls. With one of the chaperons, Dr. Boyd, he drove to Glenwood to view the lignite beds. He was told that tests showed the lignite to be about thirty feet thick. "If this bed should possess half or a quarter of the thickness ascribed to it, it will be a wonder," he wrote in his journal.

He took the train from Pink to Geneva where he stayed with his friend Ellsberry, and they "talked University till about 11 o'clock." Next morning he went out with Mr. McKinnon "behind his pair of Texas ponies and we have ridden about over the country more than thirty miles . . . As we were going down a badly washed descent to a branch the ponies broke into a run which for a few minutes seemed to make certain destruction of the buggy and its occupants, for the road was a few yards beyond where Mr. McKinnon stopped the runaways, absolutely impassible and we should all, horses, men and buggy have been precipitated into a gully. We ran into one stump which fortunately had a layer of rotten wood on outside and this was torn away by the wheel or we should have been thrown out here. Many spokes of both front and rear wheels on one side were broken, but we could still keep up and drive back. As we could go no further along this road we lifted the buggy around and hitching the ponies continued to drive back . . ."

By July 9 he was in Goodwater looking at graphite with Mr. Bunting who was interested in opening a deposit there. "After lunch we drove out past Mt. Olive to another locality where a heavy ledge of very fine graphitic rock occurs for a distance of a mile. The graphite is of exceptionally good quality." On July 10, he came to Talladega aboard the L & N. "I was met by Dr. Hal Johnson [head of the Talladega Institute for the Deaf, Dumb, and Blind] and went with him to the School. He is doing a great deal of building and improvement on the place, and is doing it in a way to make one envious. The grounds also are so well kept and free from old paper and other disfigurement that characterize our campus that comparison is inevitable.

"He drove me out to [Guiseppe] Moretti's place where we found the artist at work on a statue of Father O'Reilly which is nearly finished. It is of pearl gray colored marble and will be excellent for out of doors. His studio has many models of works on which he is engaged. Among them the monument to Miss Cahalan which is to be erected in Birmingham. The quarry is not yet really opened, only some of the surface materials have been removed and a number of blocks have been sawed and channeled out. He gets some very large sound pieces without flaws, up to twenty feet or more in length. The greater part of the rock as yet got out is of bluish cast, very fine grained, taking glass-like polish. But there are beds of white statuary marble as fine as that from below. Especially is this to be seen at an upper quarry about one fourth mile from his main works. While white like that near Sylacauga, this is closer grained marble. The prospects are that when the quarry has got into stone beyond the influence of the weather, it will be equal to any in the state and of superior quality for many purposes.

"Mr. Moretti has built a house on a hill facing the works and he and Mrs. Moretti insisted on our going up and inspecting it. He showed us first his wine cellar which is well stocked with imported beer and wines, the latter mainly Italian. The house is of very original design and all the more interesting because of the oddness. Mrs. Moretti, a Massachusetts woman, very intelligent and cultivated, was cordial in her welcome. They insisted on our staying to supper, to get an Italian supper, but we were expected back in Talladega and returned."

From Talladega, Smith went to Huntsville where he met with businessmen who were interested in developing the town. He toured the gas wells that were "in progress for gas. Unfortunately the three wells last bored have proved failures though they have gone below the depth at which the gas was reached in the others. Our drive included a visit to a cave on a hill not far from the resevoire of one of the factory companies. This cave has been a place of resort but is now abandoned as the electric lighting and other conveniences did not attract enough visitors to make it pay."

On July 13, Smith and the local men "went up on Monte Sano over many hundreds of feet of the mountain limestone. Specimens of this limestone and of the residual clay overlying, were taken for testing as to their fitness for Portland Cement. I have no doubt they can be so utilized. On Monte Sano we drove around a good deal, visiting among other things the old coal mines."

Smith returned to Tuscaloosa on July 14 and made a round trip to Chicago by July 25. Going and coming he had "the company of Landon on the train for several hours." Within a few days after returning, he joined with Mr. Hodges, Mr. Aldrich, and others in Birmingham to start a tour that included the Plantersville area of Dallas County and ended for most of the party at Waynesboro. Smith and Mr. Hodges, however, continued. He wrote: "The road to Winchester . . . [has] little or nothing of interest, generally pine woods, butchered by the turpentine man . . . Winchester does not offer many attractions to the wayfaring man."

When they reached Jackson, the two men boarded the *Russell* proceeding downriver to Mobile. While there, on August 12, Smith took time to go with Truman "down the bay where he is making readings of the tide gages with a view to correcting them." The next day Smith called on H. P. Loding in Mobile and "spent an hour or two looking at his collection of beetles (Ala.) which is quite complete. He also has a collection of reptiles and can thus assist in getting up a list of the Ala. beetles and reptiles." [Because of Smith's interest and friendship, Loding later presented "about 1000 named species of Alabama coleoptera" to the university museum, the *Birmingham News* reported February 9, 1910.] Truman spent that night with his father at the hotel, and next morning Smith saw him off "down the Bay where they expect to remain till Sat. At 10:40 Hodges and I took the M & O train and arrived at Tuscaloosa via Meridian and the AGS about 6 p.m."

Smith spent a week in Tuscaloosa catching up on his correspondence, then he was off for Talladega. He saw nothing new since the time he hiked with Landon "some years ago from about Eagle Forge up to Kent's Mill . . . After lunch we drove up by Jenkin's Springs to Mr. Franklin's where we spent the night. At the springs there was a party of young people, about a dozen who were picnicking. At the time we were there, about four, a slight rain set in which continued into the night. I was therefore not able to go down to the black cut to see if I could find a little more turquoise . . . Mr. Franklin gave me a good specimen of an Indian tomahawk of peculiar shape about five inches long. No hole is bored through it and it may be an abortive work thrown aside." He sketched it in his journal.

Smith assisted other university personnel, on the afternoon and evening of August 24, to interview prospective students in the Geneva County area. The next day, he went to Slocomb to board the dummy line for the two-mile ride to the newly opened Aus-Kel Springs, named for the two developers, Ausley and Kelly. He found the hotel with the usual resort accompaniments built near springs of pure, blue colored water boiling up through white sands. "This has been ponded by a dam at one place . . . , while a hundred yards from this several other springs boil up out of a head. I took several photos of these, also of hotel and visitors. The temperature of the water is about 67.5°."

In Dothan on August 27, Smith wrote: "Returning to Coffee Springs [from Chalky Bluff, Ruffin's Cut, Bellwood, etc] by noon train I had time to go down to the Spring and collect two bottles of the water for analysis. Time being too short for packing these up and expressing them one of the young men in A. R. Mosely's drug store kindly agreed to attend to it for me. Returned to Dothan about 5:30 to spend the night."

When Smith arrived in Montgomery, Frank Lockwood, the architect, and his family met him at the train station in their automobile. "I came to their house and after supper we had an auto ride through many of the streets, taking in some of the moving picture shows . . . Spent the 29th . . . going over plans of the museum building with Mr. Lockwood." In such a low-key manner did Smith write of the coming reality of his decades old dream of a building to house the university's collections.

Worry over Truman's upcoming wedding to Pearl clouded October for Smith. Landon wrote him that he was to be Truman's best man, but Smith understood that Truman had asked someone else. Smith was also upset because Julia had not received an invitation to the wedding: "I urge you right now to send an invitation to Julia by the very first mail, even if you have to borrow one." He reminded Truman to also invite Jones, a fraternity brother, "but above all don't forget Julia." Two days later he wrote the prospective bridegroom: "Jones has his invitation all right—but Julia's has not yet come to hand so you had better forward one at once. I can't imagine how the invitation has gone astray." On October 23, he wrote Truman: "Julia's invitation came but on enquiry I found out she had already received one at the regular time. I don't know how I came to overlook it for I must have got it out of the P.O. and given it to her." So Julia ended up with two invitations to the wedding, and as it turned out she did not even attend. There is no record of what she thought of this hurly burly. Smith sent to Mobile by express the family's gift of silver with the comment to Truman that it would " be a good start for you and can be added to from time to time."

Jennie had wanted the wedding at midweek rather than on the weekend; Smith had urged the couple to wait, to let Pearl teach a few years (she earned her M.A. degree at the university that year) while Truman continued working. Disregarding all of this, Truman and Pearl married at high noon on Saturday, November 2, 1907, at the Government Street Methodist Church.

Only Eugene and Landon (who served as best man after all) attended from the groom's family.

On November 20, "Papa" wrote Truman in Yolande, east of Tuscaloosa, where the couple had set up housekeeping, that Aldrich's wedding gift had not yet arrived and that Miss McCalley's silver ladle could not be exchanged because it was monogrammed, but perhaps the silver ladle sent by Taylor Palfrey could be. Truman complained to "Papa" that his work with the mines kept him away from home overnight. "Papa's" reply: "You knew perfectly well what engineering work called for and I expect Pearl did also." Smith continued to suggest that Pearl teach to help out financially.

In less than a month, a powerful explosion in the lower depths of the Yolande Mine No. 1 destroyed surrounding buildings and timber for a great distance. Attempts to rescue the miners entombed were severely hampered by gases and intense heat. Eventually thirty-eight bodies were brought to the surface, nearly enough to fill the forty coffins the Yolande Coal and Coke Company officials initially ordered from the Birmingham undertaker. At least twenty-five more coffins would be needed.

Smith went immediately to the scene to see about Truman. As soon as he ascertained that his son was safe, he returned to Tuscaloosa: "I'm glad I did not stay over at Yolande last night, for I can't get over the little I saw while there," he wrote Truman on the day after the accident. "Neither Pearl nor your mama nor I would draw an easy breath hereafter if you continued the work at the mines." The next day (December 19) Smith warned Truman "there may be loose slate in some parts of the mine because of the explosion."

Eugene and Jennie's letters of 1907 were dominated by their sons' affairs with scant mention of Julia and her activities. Julia's letters to her brothers concentrated on cheering them up, comforting them in their discontent, in their money troubles and other mishaps, and giving them news of the family's numerous relatives. Nowhere was there a hint of the tragic event forthcoming in April 1908.

Waterfalls on Talladega Creek near old Eagle Catalan Forge. March 18, 1908.

1908

AUTAUGA, BALDWIN, CALHOUN, CHOCTAW, JEFFERSON, MONROE, MONTGOMERY, MARENGO, TALLADEGA, TUSCALOOSA, AND WASHINGTON COUNTIES

In January of the New Year the Tuscaloosa paper noted that the faculty committee on athletics, chaired by Dr. Eugene Allen Smith, had investigated charges that some members of the varsity football team of 1907 received pay to play. A letter signed by Dr. Smith announced that after careful investigation the committee found no evidence to support this charge.

By February, Truman had taken Pearl back to Mobile to live. Smith lamented to him: "Merrill [now a teenager] went to town last Saturday . . . and got a double barreled shot gun for $18 or so. I hope he will soon develop a conscience about killing birds and other things merely for the fun of it but I suppose this kind of feeling must come to him gradually." Smith planned to send Merrill to Mobile to be "prepared" by a tutor for entering the university. Meanwhile Truman had sent home for books on botany with the idea of studying plants. Smith packed a box of books to send him and wrote: "Pearl could get a few flowers and leaves during the day and you could study out their peculiarities and the descriptive names at your leisure."

At the end of March, Smith wrote Pearl that he and Jennie were getting a permanent cook at last. "The new cook begins in the morning, then the company is coming. We have fixed up the two rooms in the servants house for her and her husband Anthony who has been with us for a week or so . . . I enclose some postals which Julia has been making. She can get a darker print of the group of four." The postal he referred to was a photograph of Dr. Smith and his three sons. It was signed JAS [Julia Allyn Smith] and on the photo side in Julia's handwriting was the message: "I'm sorry I haven't a darker print to send, but it may be weeks before I have time to print again so I'll not wait. I wish you and Pearl could be with us while Landon is here." On the reverse side she wrote, "Anthony's wife starts work tomorrow. We are having lovely weather—even Papa's baseball rains have not come. With love, J.A.S."

On April 6, Jennie wrote Pearl how she now had "a good old man and his wife" to do all the work "and I like them both very much, and as they have moved on the place I hope to keep them a good long time. I have begun to

Julia, Smith's only daughter, made this "postal" of her father and brothers in March 1908. She addressed it to her brother Truman in Mobile.

think and take interest in other things especially my flowers. Everything is looking very beautiful. My Marechal Niel roses are perfectly lovely. I wish you all could see them. Truman was so fond of them."

Jennie's sister Rose Lewis accompanied by Rosebud Lewis and her little boy had already arrived. Much more company was expected—they had been waiting for Jennie to get a cook. "I have had to spend more time in the kitchen and have had poorer cooking I think than anytime since I have kept house." Jennie covered all the recent news, and upcoming weddings including that of Mr. Kellerman, Truman's former employer. Pearl, in turn, wrote charming letters to her in-laws including one to Julia dated April 12.

In the afternoon of April 25, Julia, an expert horsewoman, rode her horse seven miles to the bridge over Big Hurricane Creek at Ruhama [now Peterson] to pick wild flowers as she often did. She tethered her horse to a nearby tree and laid her hat on the ground with bunches of flowers she had already picked. W. A. Hinton, crossing the bridge toward Ruhama, heard a noise in the water below the bridge. He saw what appeared to be a person sinking in the creek. He ran up the long hill to the mill and raised the alarm. He and Mr. Barberee, commandeering boats, immediately began efforts to retrieve the body. The news was telephoned from Hinton's home to Cottondale and Tuscaloosa. Rescue parties in borrowed automobiles and horse-drawn carriages

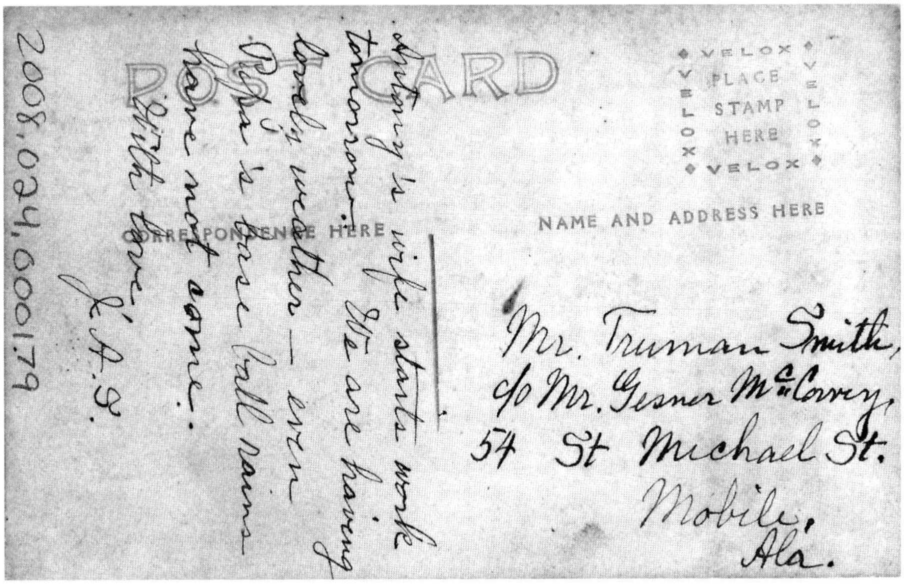

Julia wrote a message on the back of the photograph she made of her father and brothers. The card is not stamped because Dr. Smith enclosed it in a letter to Truman.

rushed to the bridge. By the time Smith arrived, Julia's body had already been recovered. She was brought back to the Smith home in one of the carriages. The recovery of her body had taken about two hours.

The local paper suggested: "She . . . [was evidently] looking for more flowers and going too near the water's edge lost her balance and fell in. The water there is probably about eight feet deep and she had sunk for the last time before Mr. Hinton could reach the spot." However, Julia was known to be a strong swimmer, and the creek was relatively placid. Speculation about her death spread through university circles and persists to this day. Suicide hardly seems to be an option considering Julia's devotion to her family (". . . all those I care for on earth."), the cheerful letters she wrote in the month before her death, her absorption in her camera work and dark room with plans for the immediate future, and, above all, the celebration of her mother's birthday the next day. More than likely, she was gathering the wild flowers to decorate the Smith house in honor of Jennie.

Julia's funeral was held at Christ Episcopal Church "which was filled to overflowing," according to the Tuscaloosa paper. "Not only the pews, but every bit of . . . standing room was occupied and many could not even get in the door. The University students attended in a body. Only a small percentage of them could obtain seats and they lined up in the side aisles, extending from the door to the chancel. The pupils of the Tuscaloosa High School, in which Miss Smith was a highly esteemed teacher, also attended in a body. The University faculty came into the church in a procession immediately after the family and bore a number of exquisite floral tributes.

"The deceased, in her life, had many friends and this and the wide friendship felt for her family . . . touched the people of this community deeply, and there were many weeping in the vast congregation, during the reading of the beautiful burial service of the Episcopal church by Rev. E. A. Penick.

"The funeral cortege from the church to Evergreen cemetery where the body was interred was one of great length and most impressive. The University and High School students formed a part of the procession, walking directly after the carriages containing the family."

The many houseguests and family members departed after the funeral except for Eugene's sister Julia, who was especially close to the young Julia, her namesake. She had taken Julia to Europe twice, most recently in June 1906, and given her a fine buggy and spirited horse the following Christmas. Now as the older Julia and Jennie put away Julia's belongings, Jennie noticed Julia's watch was missing. She asked Julia if she knew what happened to it. Aunt Julia wrote Truman in Mobile: "Of course I had to tell her that you had taken it to the jeweler to have it put in repair." Perhaps the watch was damaged when Julia drowned. Whatever happened, rich Aunt Julia redeemed the watch and billed Truman for two dollars. "Your papa and Mamma are bearing up really better than I expected they would, but of course it is very hard on them," she wrote.

The month following Julia's death Smith shoveled out the first dirt for the foundation of the University's new museum and geology building to be named Smith Hall. The *Birmingham News* noted: "It is of Ionic architecture built of Bedford stone and pressed brick" with the cost of construction set at $100,000, a far cry from the $25,000 Smith begged Henry DeBardeleben for on Christmas Day 1881.

The same newspaper reported the indignant protests of those attending the Manufacturers Association smoker in Birmingham because St. Louis brick had been bought for the new building. "That the geology building should be built of anything but Alabama material is farcical," one man was quoted as saying. The item added, "The state geologist urges that Alabama material be used."

Ground breaking for Dr. Smith's long-dreamed of museum; Smith is second from left. Although the cornerstone was laid in June 1907, the building contractor was not hired until March 1908. Photographer unknown.

In June, Smith went to Montgomery to work on the plans for the new building. Aldrich joined him there, and they continued to Troy by train. Mr. Bashensky drove them in his automobile to Absolom Jackson's lignite works. As far as Smith could see, nothing had changed from their previous visit "so . . . we know just what we have known before and no more." They had dinner with Mr. Jackson then returned to Mr. Bashensky's where they examined the ravines back of his palatial home.

Smith was back in Montgomery in mid-July conferring at the Commercial Club with officials of the state fair concerning the university's exhibit. From there he took the train to Loxley to visit Mr. Middleton who had a beautiful spring near his house. The spring "boils up through white sand at the edge of the swamp or slash. If we take up a handful of the white sand, brought up by this spring, it is nearly fluid when held in open hand and gradually flows off, but if squeezed it gets dry and hard and crushes together with a singing sound. Opening the hand the sand immediately liquefies again and this may be repeated at pleasure. I filled a preserve jar with this sand and water and apparently half sand and half water, screwed on the top and carried it to the house. Discovering a crack or weak spot in the jar I transferred the sand to another. On opening, the sand was solid and dry and did not pour out on turning the jar upside down. I had to scrape it out with a spoon and as thus taken up it was hard and firm and apparently nearly dry. I could not pack it into the other jar until I poured in more water and churned it all up for a time. This is certainly the most curious behavior I have ever noticed. Brought the jar along for further examination. I have tried since a number of wet sands but none were at all like this."

Mr. Middleton went with Smith to spend the night at Mr. Etheridge's farm situated "on the high flat red table lands with ponds, most beautiful lands, good crops of corn, cotton, etc." They came down to Brownley where a few houses remained and "a great three story factory building, built ten or fifteen years ago—painted yellow. They say this was never used at all—no machinery ever put into it. Across Bay Minette is the brick wall with some pillars, around the old Sibley dwelling place, now grown up with cedars, pines . . . 6"-8" in diameter. One old chimney still standing."

They then came "to the flat lands of Blakeley" where Captain Smith had been stationed with the Alabama Cadet Corps in 1864. "A few . . . houses are now to be found on the outskirts of old Blakeley, but of the main street of the town, not a house is standing of those which lined it in 1864, but the grand old live oaks are still there, draped with long moss."

Up the Tensaw River from Blakeley, Smith toured a fuller's earth company, "a well appointed plant with engine, pug mills, vats, drying arrangements and presses. The clay is taken from a bluff on the river and from the extension of one of the beds below water level, whence most recently it has been brought up by a steam shovel." After looking at another site or two, Smith caught the train to Mobile where he spent Saturday afternoon to Sunday with Truman and Pearl. A letter from Jennie reached him there: "Merrill is sick, just as you left him, the doctor said he should stay in bed four or five more days, not even sit up in a chair." She was doing all she could to help him bear his confinement, she wrote, then added, "I suppose you are happy . . . and getting plenty of figs. Please let no one see my letter."

Smith met Mr. Hodges at the hotel Sunday night, and they went by train to Cullom Springs. They went directly to Willie Middleton's "where I got a sample of some very foul smelling water from a well in his yard, the surrounding rocks are Buhrstone, the water is called bitter water. We next rode down to Bass' Salt Pond which is a bare spot in a palmetto swamp, caused by salt ooze, this is several acres in extent and is on a large scale what we see about many of the salt wells over in Clarke Co.

"From this place we were misdirected and came out too far west so that we came into Bladon by the St. Stephens road leaving buggy and coat, etc. at . . . [Willie] Middleton's to be sent for today [July 22]. It began to rain about 10 o'clock and we got pretty wet and because of continual showers we kept wet till night. Not much accomplished. Hodges remained behind to make some field assays of some waters about Cullom."

Smith and Hodges moved on to Bladon Springs to fill bottles "with the . . . sulphur water and packed them up ready for shipment, hoping that Mr. Turner will send them to the RR by this time next year." Near Healing Springs they investigated what the locals called a "suck" on Turkey Creek. "This is a sort of shaking bog or part of a swamp grown up with cat tails. At numerous points in this suck the whitish crust is puffed up, by some gas pressure, I suppose, in little hillocks which crack open and out of which oozes a bluish semiliquid mud. Some of these mud lumps are of pretty good size, several feet across, others are only a foot or two in dia. I suppose the suck will occupy perhaps an acre, but this is a mere guess as I did not investigate an eighth of an acre. Where the mud eruption takes place the material below seems to be of the nature of a quagmire, and rails may be pushed down into them, their own length. At one place in this suck they have sunk a barrel and provided a spring which gives a water pretty strongly impregnated with . . . [minerals]."

On August 14 "After leaving the Anniston-Alexandria road . . . passing Morrisville on Cane Creek . . . a most unexpected sight was here, viz. a large frame building used as a foundry shed and another occupied by drills, lathes and other heavy machinery for working iron objects. The floor of this bldg. was about half gone, great holes in it, yet here were the machines, held up in some way and still in use. Both bldgs appeared to be about to fall in. This is the plant of the Morrisville Machine Co."

They followed the road to Cane Creek P.O. and down to Sulphur Spring where they stopped for lunch. "The Spring is in a way protected by a large tile drain pipe, but the horse, cow and goat lot is immediately above it and a kind of lot drainage bog is in the lower part of this lot going up to the Sulphur

Spring, close enough to temper one's enthusiasm."

After the close of the 1908 summer tour, the *Montgomery Advertiser* (September 9, 1908) published a personal close up of Dr. Smith: "His brown eyes . . . are youthful because they have never been clouded by doubts regarding the thing that they were placed in the world to see and to accomplish . . . His voice is gentle in the extreme . . . His manner of speaking is peculiar, he seldom closes his lips at the end of a sentence but holds, as it were, a suspended smile through all his conversation . . . His frame is light—a 16 yr. old son [Merrill] stands head and shoulders above him. He is spare almost to leanness. He cares but little about dress . . . For 35 yrs he has laid the foundations of riches for others. In making many rich he has yet remained a man on a professional salary. The prize of his high calling has been 'Service to Alabama' and no emolument other than this has ever appealed to him. He had been tempted with offers from other States, but his self-made pledge to give one whole life to one great task has held unbroken.

". . . A bed of cretaceous fossils in Autauga Co. first interested him as a boy. Some of these fossils and a number of Indian relics gathered by him before he was a man now form a part of the collection of the State . . . He loves visiting and conversation in good company. He loves novels as a relief from the tension of his work; he loves travel—through books. He loves Scott, Dickens, Dumas. Dickens he has read and re-read everything and then wanted more. He loves and honors Burns."

The *Tuscaloosa Times-Gazette* (October 8, 1908) gave notice of a geologist's dream trip arranged by Dr. Eugene A. Smith. A group of distinguished scientists, aboard a houseboat towed by a launch, were to descend the Black Warrior and the Tombigbee rivers to Mobile, then ascend the Alabama River to Camden, there to take the train for the return to Tuscaloosa. "The trip is bound to be one of great interest and of great advantage to science. There are a great many formations, the naming of which differ in different states, and their journey will tend largely to settle these differences and bring about [a] general system of nomenclature that will save much confusion."

A member of the party, the Reverend Mr. H. E. Wheeler of Blocton; listed his fellow travelers: "Dr. T. Wayland Vaughan, U.S. Geological Survey; Mr. Earle Sloan, State Geologist of South Carolina; Mr. T. W. McCallie, State Geologist of Georgia; Dr. E. H. Sellards, State Geologist of Florida; Dr. Eugene

A. Smith, State Geologist of Alabama, and his assistant, Dr. F. W. Prouty; Dr. George Little, formerly professor of geology at the University of Mississippi; Hon. T. H. Aldrich, Paleontologist of Birmingham; Dr. R. M. Harper, of New York, Botanist. For part of the trip we were honored with the company of Mr. Chas. Butts, of the U.S. Geological Survey."

After the trip, the Reverend Mr. Wheeler wrote ". . . the water being low and the weather fine, we had exceptional opportunities for studying the exposures of cretaceous and tertiary stratas on our course. Our botanist was busy listing our rich and interesting flora, and discovering rare plants in unusual localities. The geologists were occupied at every bluff, taking stratigraphical notes and measurements, and the whole party assisted in the pleasurable work of collecting the fossils of successive horizons. Several cameras aboard were brought into frequent requisition to catch some beautiful bit of scenery or some striking geological formation. Our cabin served both as dining room, sleeping apartment and library, for the literature of the Coastal Plain was fairly well represented in the books which the several members of the survey had brought for reference. Our evenings were devoted to the discussion of problems, geologically interesting or otherwise, and to the transcription of notes. It was a busy time after every fresh draught of fossil, the preparation of lists and identification of species being in the nature of a round-table conference . . .

"The shoals between Demopolis and Naheola added spice to our other experiences, and more than once the problems of navigation were the serious concern of our geologic minds, requiring in two memorable instances our practical services overboard as a sufficient inducement to our launch to follow a more congenial channel. But the scenery along the route, especially the lower Tombigbee is worthy a special trip. Different geological formations succeed one another, fine examples of cross-bedding and step faulting being observed, indurated ledges and concretionary masses of fantastic form appearing now and then, while the river banks were enriched with a varied vegetation, and birds of strangely different form and plumage appeared continually on the scene. Conspicuously beautiful are the white bluffs at Demopolis exposing in all its glory the 'Selma Chalk' formation, which has proven of great value to the manufacturer of cement. At Moscow Bluff, near which we also had a trying experience with shallow water, we found the actual contact of cretaceous and Eocene beds, and a geological freak showing a mix-up of both formations with the fossils peculiar to each in the strata of the other.

". . . Our 'outfit' was the first to pass through the newly constructed lock at Demopolis, where we were shown many kindnesses by the government engineers, and with the completion of this and one other lock, a channel from Tuscaloosa to the gulf will be assured.

". . . The white limestone beds which overtop the Claiborne formation here are interesting for several reasons. To the paleontologist for the fossils they contain, to the housebuilder, since the soft rock easily shaped with saw and plane, weathers into a most desirable chimney stone, in general, because in the prairie hills of Clarke and Choctaw counties this formation has yielded the enormous bones of the gigantic Zeuglodon, a complete skeleton of which was found by Harlan in 1834."

1909

JEFFERSON, MOBILE, TUSCALOOSA, AND WASHINGTON COUNTIES

In mid-January of the new year, Smith was concerned over the drops of grease from the workers' pulleys that had stained the porous blocks of stone for his new building and "shows pretty badly," he wrote Truman. He had tried, with no success, the solutions he knew. He asked Truman to "look among your recipes" to find a remedy.

This year marked the twentieth since Daniel W. Langdon left the university. He tried several commercial ventures before locating in New York where he eventually became a landscape architect. He married Berenice Frances Webster, a well-known sculptor who had been a student of St. Gaudens and Rodin. In March 1909, Mrs. Langton came to Tuscaloosa to make a life-size bust of Dr. Smith. The work of art was to be a gift from the Langtons to the new Smith Hall. (Langdon now spelled his name with a "t" instead of a "d" because he had discovered the former was the original spelling of the family name.)

Smith wrote Truman on February 12, 1909: "Mrs. Langton and her little girl arrived from N.Y. last night . . . We will fix up the Geological room for her workshop which will make a very good one with the aid of a stove or some sort of heating arrangement during cold spells.

"All of the columns are up and capped so that the front of the new bldg is

Smith Hall construction, February 1909. The building was ready for use in the fall of 1909.

looking all right. We are having, however too much rain for the best progress of the work." In a letter dated February 20, Smith asked Truman to find two bottles of Dr. Mohr's cologne for Jennie. "Your mama don't want any other kind." Mohr, eight years dead, had named his cologne "Farina" and on one occasion had sent a half-gallon of it to Judge G. W. Clinton in Buffalo, New York, who also fancied it. Smith continued: "Mrs. Langton is working on the bust which is all blocked out but will of course take some time yet to bring out the details."

Later in the week, Jennie wrote Truman her version of the Langton visit: "I suppose your papa or Merrill has told you about what a gay bird we have in the little Langton child. She keeps things in a stir I can tell you. She dresses as a boy and is as lively as any you ever saw. We all like her though and I think she amuses your papa a great deal. Mrs. Langton is very entertaining. She has traveled a great deal, and knows so many interesting people such as artists, musicians, writers, painters, etc. She is not very swell in her appearance, but

Berenice Frances Webster Langton created this bust of Dr. Smith, which was to be a gift from the Langtons to the university. The bust's whereabouts are unknown. Photographer unknown.

we forget that when we listen to her talk."

On March 24, Smith broke the news to Truman that Bob, the family dog, veteran of many summer geological trips, had disappeared. "He was in the yard Sunday afternoon and we have not seen or heard of him since and this is Wednesday evening. I have walked all about here with in a radius of a mile but can find no trace of him. I am sure he is dead, for you know that he never ran off from the place for longer than a few hours. We are all very much grieved about it, though in the nature of things he would not have lived many years longer." In a later letter, Smith wrote, "I have had enquiries made in every direction." Finally, on April 2 he wrote Truman: "I found Bob's body down in that . . . settlement which is building up at the mouth of the main avenue going up to the Asylum from the Huntsville Road. I think he must have been killed by dogs which are pretty bad down there. We brought the body up and buried it in the corner of the yard near to where Frisky was buried."

After seven weeks, Mrs. Langton departed with the clay model of the bust—the photograph of it in an unidentified newspaper showed a frowning life-size head and shoulders of a clearly aging man. It was intended that the bust be cast in bronze by Walthausen, "perhaps the best known maker of casts in the country." The reporter added, "In spare moments during her stay here Mrs. Langton did an exquisite full length portrait in low relief of her little daughter, Berenice."

In mid-April, Smith confided to Truman: "I was a little under the weather yesterday and last night but am feeling some better this morning though not the best in the world. I think I had a sort of dumb chill." On April 27, he reported to Truman: "Mr. Hodges took a picture of the new bldg a few days ago, showing its present condition. You will notice that the two wings are complete and they are putting a tin roof on right now. The central part will have to run up about ten feet higher . . . These tremendous rains are very bad for the bldg because so much water accumulates in the basement. I am afraid it will soften the clay foundation and cause settling and cracks, though nothing of the sort can be seen yet. In a little while after we trim up some trees so as to get a view from a little further off, I will send you another picture showing more."

To Smith, commencement loomed large in the school calendar. In his early years on the faculty, he had invited friends from his undergraduate years to be his and Jennie's houseguests. That custom had gradually faded, and now the guests were mostly family members. Smith always took an active part in the planning for the celebration, serving on one or the other of the university committees. This year Jennie invited Pearl and her friend to come for commencement. Smith wrote Truman that the two girls "could come together and stay together in a room so that Pearl might not be afraid."

On May 30, Smith wrote Truman about the plan "to remodel Woods Hall at an expense of $30,000 or more, will be tearing off all the plastering, taking up floors, and putting in new partitions immediately beginning in the morning." Smith had begun his tenure at the university in the multipurpose Woods Hall, the first building to be erected after the Civil War. A June letter to Truman told news that saddened Smith "I just heard . . . Saturday evening of the death of Mr. Langton who committed suicide by drinking cyanide of potassium . . . It has been a great shock to me for I thought a great deal of him and in some respects he was one of my most intimate friends."

An item in the *Tuscaloosa Times-Gazette* (July 1, 1909), reprinted from the *New York Times* (June 22, 1909) gave more details: "Daniel W. Langton, a landscape artist with offices at 341 Fifth Avenue, New York, committed suicide last night at his home at Hoyt's Corner, three miles from . . . [Morristown]. Dr. George L. Johnson, a Coroner, was called on the telephone at 11 P.M. and went to the Langton home. He found the man dead in a chair, in the sitting room with an empty bottle in one hand.

"The physician was informed, so he says, that Mr. Langton had taken poison about 7:30 o'clock. The bottle, he said, contained cyanide of potassium. No information could be obtained at the man's home, an effort being made to keep the suicide secret.

"The last seen of the man by his neighbors was about 6 P.M., when he came home in a hack. He had been to New York all day, returning to Morristown on the train reaching here at 5:20 o'clock. He appeared to be in good spirits at the time, saluting several of his acquaintances on the road home.

"Mr. Langton was 45 years old and had lived in this vicinity for ten years. He leaves a wife and two children." Smith's tribute to Langton in the *Bulletin of the Geological Society of America* (May 21, 1910) praised Langton and his wife as brilliant talkers who made their New Jersey home a gathering place for writers and artists. "With these surroundings, however, he never lost his early love for geology," Smith continued, "and he was often heard to wish that he could afford to go back to the old delightful occupation."

Smith wrote to Truman on July 21: "The summer school ended yesterday and the place is a little lonely now, but all the better for that. The work is going on about as fast as we could well expect though it does seem to me to drag in the museum bldg. On the roof of the main part they lack still the glass skylight, the rest of it and in fact all that we can see from the outside below, looks finished. They have just received the sash and doors and frames so that these will be going in from now on. I foresee though that I will hardly be able to move in much before Christmas unless the woodworkers get a move on themselves. They have most of the cement or concrete floor of the museum hall in, all except that over the main central entrance hall.

"Mr. Jake Marks who is drilling for oil over in Baldwin, has been writing me and I may run down to Foley some time soon. I don't know that I can get over to Mobile at that time, for I don't like to be away from work long at a time, for something is always turning up that needs looking after. Thus today I found that the risers for the radiators in the museum were brought up right in the space where the exhibition cases are to be and they had to be changed, even at the expense of cutting through the concrete floor.

"We have been doing very well lately for figs, and I have had as many as I wanted several times. I don't suppose I shall get any of the Mobile ones as they will be gone before I come." Merrill's friend Morgan was spending the night with him. For some weeks the two boys had been building a boat. "I am afraid they will make a poor job of it but it may satisfy them."

Later in July, Smith assured Truman again, "We are getting the greatest plenty of figs here now from the McCorveys, the Gorgases and from a bush at the stable so that missing them in Mobile will not count as much." A heavy rain was falling August 9 when Smith arrived in Mobile. "Went down to the Chickasaw and took dinner with Truman, after which we spent the afternoon in arranging about sacks etc for the Coal Bluff lignite to be sent up on Tuesday by the City of Mobile. Spent Sat. night and Sun. with Truman."

Smith wrote to Truman on August 26: "I don't know whether Merrill takes any special interest in Geological or rock formations, but our going on top of Red Mtn gave him a very good chance to see the great B'ham Valley with its dozens of sml towns fringing B'ham. I think he enjoyed the trip to East Lake, where he took a ride on the Scenic railway and also on the human roulette wheel, the latter was right funny. It is a disk revolving on a vertical axis. The people seat themselves flat on it and when the speed increases sufficiently they are slung off in a very ludicrous way the floor outside being level with the outside edge of the wheel."

Smith had hopes in early September that the serving city engineer of Jackson, Mississippi, would give up his position because, he wrote Truman, "Landon would stand a good chance of succeeding him, and that would suit him more than anything else, as he is in love with a girl at Jackson, and if he were stationed there it would save him a lot of money which he pays out every week in going down to Jackson and staying over Sunday."

Smith continued with his usual update on the museum telling Truman everything had been held up because the glass for the skylight had gone astray, and they could not trace it. "They are putting all the hands they can get to work on Woods Hall trying to have it ready for the opening. I think it will be in such shape as that the students can move at once into their rooms but there will be a good deal still to be done as to the outside finish, stuccoing etc. before the job will be complete. We have only about a week now before the opening . . . The president seems to expect about the usual number or a slight increase. A large increase is hardly to be expected in view of the raise in the entrance requirements."

Smith sent another update on the museum to Truman on October 8: "It seems to me that it takes about as long to get your electric machinery installed in the boat as it does Mr. Bynum to finish up my bldg. The show cases have come and are stacked up in all the halls downstairs as there is no room upstairs. They are getting along fairly well putting in the ornamental work in the museum hall. They have the capitals on many of the columns and have

Dr. Smith anxiously watched each step in the building of the museum and kept his sons informed. He made this photo of the completed Smith Hall in 1911.

the cornice pretty well up. All this ornamental work is going to be splendid, and there will be no such handsome hall in the country that I know anything about. The cases are very handsome of oak of same style and finish as the woodwork in the bldg with the drawer pulls, etc. to match the door knobs, locks and other hardware."

Smith's plan to send Merrill to Mobile to be readied for entering the university had become a reality. Jennie received a letter from their youngest son the day after he arrived; Smith, who hoped Merrill's studies would enable him to enter the university without conditions, heard from him a bit later. "Do you think Merrill has been homesick?" Smith asked Truman. "I was afraid he might be." And he signed this letter of concern in his usual way: "Affectionately, Eugene Allen Smith."

Out in the field again, Smith and his party were in Vinegar Bend, Alabama, on November 10, having left Buckatunna, Mississippi, the previous day. They took a southeast slant to State Line where they arranged to drive out about eight miles to see a section they had heard about at Mr. Avera's. They looked around the place, and Smith made notes for his journal until noon when he asked Avera about taking dinner with him. But "he declined, so we returned to State Line where we got something to eat at the hotel and heard a good many anecdotes of Mr. Avera from the landlady—all going to show how close and unaccommodating he is. One of these stories relates to an old negro man who had been Mr. Avera's father's slave, and who had worked for the family all his life. When this old negro died and was to be buried on east side of river, Mr. Avera charged ferriage for the corpse and collected it from the more humane friends of the old . . . [man]." From State Line they continued angling toward Mobile where on November 11 Smith checked on Merrill and the two of them got together with Truman.

Merrill's letter of December 4 showed not a trace of homesickness: "Emmett and I were riding around with Dr. Frazer in his automobile most of yesterday afternoon. Emmett has almost learned how to run it." Smith warned Merrill to be careful about driving the car—he might run into someone or a ditch. He suggested that Merrill drive only with some family member in the car with him.

1910

Tuscaloosa

The exciting but overwhelming final days of the move into Smith Hall were well chronicled in Smith's letters to his sons. "We are right in the midst now of moving things over to the new bldg from the office and from the room upstairs," he wrote Truman on January 13, "and are certainly in a state of confusion. I don't know how we are going to manage the office in the interval before we get the shelves up in the new quarters."

To Truman on March 7: "The weather is fine and the boys are making the best of it in their ball practice. We have our mid-term today and tomorrow so that some of them are staying in though the field seems to be pretty full. I can see the field perfectly from the window of my new office where I now am. It is a little lonesome over here as you may imagine, but the office is a delightful room. I have not yet put up book cases and other furniture so that every thing is in some confusion. Still I am beginning to feel at home here.

Dr. Smith's library in his home on campus. He later donated his books and pamphlets to form the nucleus of the university's geological library. Photographer unknown.

When we get the gas and water connections and Hodges gets to work things will be more lively."

On April 2 to Truman: "We have about concluded to undertake the moving of my library over to the new bldg before Commencement. This will mean work. At the same time we want to have by Commencement a good display in each of the new cases, 16 in number, which also means work. Still when I remember that Mr. Langton and myself put all the specimens on the shelves in the cases in the old museum one Sunday, I see the thing is not impossible. Still I don't propose if it can be helped to work this time under such a strain."

To Landon on April 28: "Commencement will be upon us before we know it. I have such a lot to do in the Museum before that time that I am often appalled, but I hope it will all work out well in the end. We are going to have a magnificent exhibit especially of shells which Mr. Aldrich has given us. The books that I have turned over to the Survey library in this building will nearly fill all the shelves, and will make a fine nucleus for a geological library." This "nucleus" that Smith donated consisted of three thousand bound volumes, eight hundred unbound volumes, and many thousand pamphlets. Eugene Allen Smith Hall was dedicated at commencement on May 5, 1910. Among many other gifts to the museum in Smith's honor was a portrait of Henry McCalley given by his sisters Kate and Carrie McCalley of Huntsville.

On May 24, Smith sent money to Merrill in Mobile to settle his debts and pay his way home. In less than two weeks, he had a job in Holt, working and rooming with some other college students. He took his parents' phonograph with him "which is all right as your mama and I have not used it very much of late," Smith wrote Truman. To Landon on June 10: "Merrill got back [from Holt] Saturday afternoon and has been to two or three parties already and has company a good deal of the time although Morgan Stickney [the friend he built the boat with] is up the river on a camp and Charlie Barnwell is sick in the hospital with typhoid fever. He seems to be improving though and we all hope he will be up in a week or so but you know how treacherous this fever is.

"Summer school opened yesterday but as it has been raining most of the time since and I have been a little under the weather I don't know how many have come in." On July 19, Smith caught the four p.m. train for Calera with the intention of going on to Fayette County's Gas Wells region. In answer to a question from Truman, he wrote: "I know the bluff . . . on [the] Chickasawhay. It is called Shell Bluff and I have been there half a dozen times with Mr. Aldrich, Judge Johnson, Mr. Wayland Vaughan and a lot of others. It is a well known spot. I have a number of big oyster shells from there here in the Museum. I am particularly anxious to know something of the bluffs . . . on [the] Pascagoula river. We came up the river once on the east side and I went down to the banks at several points, but saw nothing of importance. That was the year you drove the mules."

In early August, Merrill was off to Prattville where he and his numerous cousins spent the days attending barbecues, visiting, and having a generally good time. He mailed a picture postcard of the waterfalls on Autauga Creek to his father in Troy, care of Mr. Carl Tutwiler.

Truman's boss came to Tuscaloosa in October. Truman wrote his parents "if possible" to have him "out home" for a visit or a meal. Jennie and Eugene invited him for noon dinner; he stayed till four. "We were very glad to get acquainted with him," Smith wrote Truman. "He made himself very pleasant and your mama was agreeably surprised as she had the idea that he was rather grumpy."

Merrill gained admittance to the university and also made the football team. On November 21, Smith wrote Landon in Mississippi that Alabama beat Tulane, but Merrill came out of the game with a sprained ankle. Other news included Cousin Dora (Julia Pratt's daughter) coming to town for a visit. Jennie and Eugene took her with Miss Mamie Gorgas and Richard, "and four or five S.A.E. boys on a launch trip up the river to Tidewater . . . It was a very pleasant trip. We left White's Landing about three and got back about five, and they all then came in and took supper with us."

Henry Fairchild DeBardeleben died December 6. A native of Alabama, at age sixteen he became the ward of Daniel Pratt. He lived with the Pratt family in their imposing mansion in Prattville and assumed responsible positions in Daniel Pratt's industrial empire. He was later made manager of Pratt's iron interests in Birmingham and from there, after Pratt's death in 1873, he became more and more influential. Being married to Daniel Pratt's only daughter did nothing to hinder his rise to the top. Henry S. Marks wrote: "His importance to the economic development of Alabama cannot be over emphasized, for he was the most spectacular figure in the state's early industrial growth being known as 'the King of the Southern Iron World.'" But the beautiful Museum of Natural History on the campus of the University of Alabama, that has been so influential in the education of hundreds of thousands of Alabama's youth, does not bear his name.

For the Christmas holidays, Truman and Pearl sent some of their flowers from Mobile by train to Amelia Gayle Gorgas, long time family friend. On Christmas Day, Smith wrote Truman that the flowers arrived in good shape and in "good time and [Mrs. Gorgas] . . . begs me to thank you and Pearl for the remembrance." The last day of December, Smith returned from his annual geological meeting (in Pittsburgh this year) but was leaving right away for Montgomery. He wanted to get a couple of bills passed by the legislature, he wrote Truman, and he needed to be present "to keep our bills pushed along and not overlooked."

Pearl wrote the family thanking them for their Christmas gifts. To Merrill: "The handkerchief that you sent me Christmas is beautiful. I received only one and I am very glad to have it as it is a very necessary article . . . Truman is very busy with his bicycle. You may know that he is taking it apart, and it has been a tedious job, lasting several hours. I believe that he would disjoint me if he thought that he could put all the parts together correctly."

1911

Clarke County

In mid-January, Smith wrote Landon in regards to his request for more money: ". . . Your living expenses ought not to be more than $50 a month and you ought to be able to save a good deal each month to begin to straighten out your affairs . . . As I said in my previous letter all this leaves me on constant apprehension lest your drafts on me should fall when I had not the money in the bank to cover them."

Again on February 21 Landon wrote from Mississippi to tell "Papa" that he had drawn on his account for twenty-five dollars. "I know you must wonder how this comes about but I assure you it is nothing but a straight honest debt." Smith had asked for forewarning so he could adjust his finances accordingly but, Landon wrote, "The reason I haven't been able to look ahead and write you before hand is because I dislike so to have to borrow money that I don't until I see there's no other way out."

In March, Smith encouraged Landon to give up smoking. "I know very well that I have been much better off since quitting . . . Your mama will go to Montgomery to have a dress fitted but will be away only a day or two." Truman and Pearl came to visit , but Smith was in Atlanta and Montgomery for nearly a week of that time. Aunt Julia, taking "Adelaide Bell with her this time," had gone jaunting about Europe for six months, Smith wrote Landon.

On March 31, Smith wrote to Truman: "Your mama goes to Prattville next Monday to be gone most of the week. She wishes Pearl were here to take charge while she is away. She is having the usual trouble with the turkeys which persist in making nests down in the field and have no inclination to set on the eggs when they are moved up in the garden. A little later there will be the worry about the young turkeys getting wet which as you know is fatal to them. There ought to be a great deal of good to be had from eating turkeys as they are such an endless trouble to raise."

No person in the state possessed more personal knowledge of Alabama roads than Smith nor, it is safe to say, had anyone else suffered as much from their deplorable condition. When the State Highway Department was established on April 5, Smith was one of five members appointed by the governor. He used his influence to get W. S. Keller appointed state highway engineer. When he wrote Truman the news, he added: "I myself could not be . . . [in Montgomery] as the train was four hours late and the meeting was set for seven o'clock. I, however, arranged it all over the telephone."

Smith had another purpose in Montgomery: to shepherd his topographic map bill through the legislature, he explained to Truman. He felt quite anxious about it. "Any little misunderstanding or misrepresentation would suffice to kill it and there is no telling beforehand what turn things will take." On April 15 to Truman: "My bill passed the house . . . thus becoming the law as it had already passed the Senate. The bill appropriates $10,000 annually for cooperation with the U.S. Survey in the making of topographic maps and I think I will go to Washington next week to arrange details . . . I have lost a good deal of time on this month, but Dr. Prouty has taken charge of my classes which thus have not suffered by my absence."

Pearl and Truman sent strawberries and flowers by train from Mobile so that they arrived in Tuscaloosa on April 25, the anniversary of Julia's death. Next day, Jennie's birthday and also [Confederate] Memorial Day in Alabama, Smith wrote them: "The strawberries were much enjoyed by all of us as they are far sweeter than what we get here. The flowers we carried over to the cemetery and placed on the graves of our loved ones," which would have included, besides Eugene Jr. and Julia, Jennie's brother in law, university President Burwell Lewis, General Josiah Gorgas, and many other friends and relatives in Evergreen Cemetery. Perhaps they spared a rose or two for Irish-born Michael Tuomey, distinguished first state geologist, resting alone under his Celtic cross.

On May 4, Smith wrote Landon: "We had an out-of-door performance yesterday of Shakespeare's Much Ado about Nothing in the afternoon and of the Greek tragedy Electra of Eripedes at night. We all went and enjoyed it especially the Shakespeare." He was working hard preparing the museum for commencement. "I am getting things to look pretty well about this building and would have it all right if I could get a little help in the way of hauling dirt or soil to level up things for a grass plat in front. The circle with grass border will be all right by Commencement but I could not afford the coleus plants and so have put out cannas, red salvias, and white flowers of some sort which will show up fairly well though they will hardly be in full bloom by Commencement." On May 15 Landon admitted to his father ". . . you are exactly right about a man being badly handicapped when he is in debt. I realize it fully." But that realization did not cause him to change his ways.

One event during commencement that year produced a toast to Dr. Smith that so pleased him he asked for a copy. Toastmaster John A. Rogers, state senator from Gainesville, said to the assembled merrymakers: "This company knows Dr. Eugene Smith. (Cries of 'Yes, yes.' Cheers long and loud.) A man who thinks in eons and understands the language of the stones. The hoary rocks reveal to him the mysteries of creation. He reads the history of the continents from the geological decorations which line the sandy margins of their seas. Knowing the past, being familiar with the present, he can, by combining the two, prognosticate the future within the limits of some million years. He can do more with a fossil bone than did the Creator with a living rib, out of the one only a woman was made, while out of the other Dr. Smith will construct nothing less than a mastodon.

"But you don't want to fool yourselves about Dr. Smith—even for a moment, for while he is always delving among dead things he is by no means a dead one, he is as live a wire as the one that carried the message to Garcia, and when he goes out to get results the ponderous law-makers of Alabama are to him but playthings."

In June when figs ripened in Mobile, Truman wanted to express a batch to his father, but he realized they were too perishable to survive the trip. Figs would have been a solace Smith badly needed as the trustees of the university had not been able to provide the money for the topographic mapping. "All this leaves me in the greatest uncertainty . . . as I don't know when I can count on [the] money . . ." and the work "ought to be begun as soon as possible."

Eugene and Jennie visited Landon in Tupelo, Mississippi, on July 5; he showed them his engineering work on the roads. The following day, they continued to Oxford where Smith spent his time "looking at the cuts" and Jennie's friends entertained her. By July 27, however, Smith had the geology wagon on the road in Clarke County in the Buhrstone hills. "On much of this hilly land the virgin pine timber is still standing. It reminded me of the old days when pine lands were not so rare as at present. The main object of the morning's trip was to collect samples of the siliceous clays with fine conchoidal fracture that has been pronounced good fullers earth. When this material is dry it is very absorbent and sticks tightly to the tongue . . . If its bleaching property should turn out to be as reported it will be worth while." Over time fuller's earth has had many uses—as a catalyst, as filters, in the manufacture of woolen cloth and talcum powders, to name a few.

Smith's traveling companion came down with a cold "and so we gave out the proposed all day trip to Tallahatta Springs and the Hatchetigbee clay area with concretions which I observed many years ago and which I was very anxious to see again . . . I am told that the Tallahatta Springs is no longer to be seen as the water escapes into the creek. I was sorry for this also as I wished to get some of the water for analysis."

In the latter part of August, Smith managed to visit both Truman in Mobile and Landon in Tupelo. Landon had competed in a marathon swim in Memphis, the *Tuscaloosa News* reported September 16. Fifteen of the best swimmers in the South entered the ten-mile race in the Mississippi River. Landon was one of the ten who stayed the course. He won a bronze medal.

The *News* described him as "one of the best swimmers that ever breasted the waters of the Warrior. He always seemed as much at home in the water as on land and seemed never to tire from swimming."

The year moved on to October when seventieth birthday congratulations inundated Smith. Truman Aldrich sent him a gold nugget that he got in Alaska "back of Dawson," and a special rhyme:

> A certain wise geologist
> His days in delving spent;
> And while he dug up many things,
> Could not dig up the rent.

Greetings came from J. J. Mayfield of the Supreme Court of Alabama: "You have passed the 70th mile post on the road of Life's Journey, with that same quick step, merry whistle, and swift-twirling cane with which (to the delight of the student body) you have for the last forty years crossed and recrossed the campus of the U. of A. In my imagination I can see you today, striding across the campus, from your house to your section room, twirling your small cane and blithely humming or whistling—just as you did when I first saw you to know you, on the 18th of September 1883. You seem to me not to be a day older in mind, spirit, or movement than on that well-remembered occasion . . . You are generally regarded as part of the institution—that without 'Little Dock Smith' there would be no U. of A . . . In the summer of 1897, I met for the first and last time that greatest modern Geologist, Dr. Joseph Leconte, of Calif., and he on that occasion said to me, 'You people of Alabama ought to be proud of Dr. Smith, he is one of the very few men whose opinions and reports I am willing to write in my textbooks without verifying.'"

Smith had a particular interest in how Alabamians used native stone. Chimneys were especially good showcases for local stonework. He photographed this chimney near Hackleburg, Marion County, November 16, 1911.

On November 11, Smith instructed Truman, a married man of twenty-five earning his own living: "I notice you spell two words wrong, viz., interfere and surprise and not supprise and interfear as you have it. Make a note of the two words." Four days later, Smith wrote the latest news to Pearl: "The electric lights have been put in the house and are appreciated except that they are turned off at 11 o'clock which makes it necessary to keep the lamps going also." Toward the end of the month Jennie, with the Christmas holidays in mind, sent Pearl a note: ". . . So we have begun to count the weeks if not the days when we will all be together again. The time cannot come too soon for me."

The family get-together at Christmas left Jennie bereft. She wrote Eugene, in Washington, D.C., for his annual post-Christmas meeting, an emotional letter telling how lonely she was after the family left her: "It was almost like a death in the family in the quiet and great loneliness of the house. I scarcely slept the night you left for thinking of you and fearing you would get sick and last night it was little better for my heart was aching so for all of my loved ones. Merrill was out until after 12 and I was all alone and will be for the next two nights as there are two more dances. It just poured down raining all day yesterday . . . They [Truman, Pearl, and Landon] all went down [to the depot] in the auto together. They would have walked for the economy of it if it had not been raining so hard for they all seemed short of funds so I paid for the auto . . . We were all so proud of your looks in the derby. I can see you now as you looked when I last saw you. Why did none of us think of a derby before, it looked so much more up to date . . . I do hope you took no more cold and that you will have a pleasant time and come back safely to me who loves you so dearly. With heart full of love, Jennie." The letter, however, did not reach Eugene in Washington. It was returned to him in Tuscaloosa.

1912

BALDWIN, COVINGTON, AND MOBILE COUNTIES

In February Pearl sent Jennie a jabot (as Truman called it) she had made. To Smith it was a 'collar' when he wrote that it had arrived. In the same letter, he complained that the north wing of the new building was not 'habitable' because of the cold. "I do not think it gets a fair proportion of the heat."

Smith wrote to Truman on February 15: "Your aunt Julia will be up to make us a two week visit next Monday. Merrill is going to a masquerade party out beyond the M & O depot tomorrow night at Henry Alston's and there will be another mask ball in town on the evening of Mardi Gras. He goes to everything of the sort that comes along. Three dances on successive nights last week."

In early March, Eugene and Jennie entertained the new president of the university, Dr. George H. Denny and his wife, at six o'clock dinner along with the McCorveys and the Barnwells. Dr. Denny came to Alabama from the presidency of Washington and Lee University in Virginia.

At this time, Truman was still on the *Chickasaw* taking soundings in the harbor among other things. He was to go down the coast to Gulfport, as there was a tide gauge there to be fixed. Money matters dominated many of the letters exchanged by Smith and his sons during this period, exceeding even news of the University of Alabama baseball team and football events. Smith asked Truman if Landon ever paid back the fifteen dollars he borrowed. If not, Smith offered to remind the older brother of the oversight, because he recently got a twelve dollar and fifty cent raise that added up to one hundred fifty dollars a month he earned from the gravel and road job, and fifty dollars from a job he held with the town of Aberdeen. Both sons, but particularly Landon, were a constant drain on Smith's financial resources. In March, he was struggling with the annual problem of how to pay his insurance, and asking Landon to repay some of the money that he had borrowed to help with the payment.

Smith wrote to Landon on May 6: "I would never disappoint a negro to whom I owed anything, nor would I disappoint any white man who was dependent on what I owed him for his immediate wants. I know you and Truman and Merrill have all that same feeling. It may keep you poor but you will not regret it in the end . . . Of course I shall not say anything to anyone outside of the family about our affairs. The matter of the notes in the bank however is naturally known to those in the bank." Smith inserted among the financial advice the news that Truman "has just sent me a plan of his new house into which he moved on 1st of May."

Smith wrote Landon about his intention to go to Birmingham (May 22): "I shall . . . get the portrait of Prof. Tuomey which has just been painted for the Museum. Mr. J. H. Fitts has paid for the painting. He is very low now with cancer of the throat and will hardly get well, his death is only a question of a short time." Fitts's obituary, published in the *Tuscaloosa News* on May 31, said in part: "In 1896 when the enemies of the University were abusing its every act, and the institution was in sore straits, it was to him more than any other man that credit was due for bringing order out of chaos and restoring the University to its position in the educational world."

A resolution upon the passing of "the University's best friend" was signed by George H. Denny, Eugene A. Smith, and T. C. McCorvey: "It was largely through his financial support that the restoration of the buildings of the University, destroyed 1865, was begun . . . At every crisis in its later history he was always ready to come to the rescue with his zeal, wisdom, and his financial aid."

The close of the school year also brought the end to Professor Smith's classroom teaching at his request. He wrote Landon on June 4: "As matters now stand I am head of the Department of Geology without necessity of teaching, am Director of the Museum and State Geologist [$2400 per annum] . . . . For now I only get $1000 and the house from the University, at the most $1600 counting the house rent at $600 ." In addition, he had agreed to carry out special administrative and executive duties as requested by the university president. He had tried for a Carnegie pension without success, because the university was not on the admitted list of that foundation. He explained to Landon, whose financial situation had become even more desperate than usual, that he himself was so pinched for funds that he thought Jennie could not make her proposed visit to friends in Tennessee. He tucked in a bit of news that could not have made Landon feel better: Aunt Julia had given "$10,000 or more" to "the new Girls' Methodist School in Montgomery" for a building—a dormitory, Smith thought—to be named for her. Julia's gift built the Julia A. Pratt Memorial Hall, first residence hall at what afterwards became Huntingdon College.

Later in June, Smith began crating up Julia's piano, her sewing machine, a music cabinet, and several chairs including the large one from Julia's room, to send Truman and Pearl to help furnish the new house. Unfortunately the project was delayed because the packing crate was left out in the rain. It had to be thoroughly dried out before he could proceed.

On June 11 at Orange Beach, Smith took samples of the black sand. "The color is due to rounded grains of some black, brown, and reddish minerals—probably tourmaline, limonite, and garnet but microscopic tests should decide this. Both at Orange Beach and along the outer beach there were places where the sands give out musical sound when the foot scrapes over it, and when walked over."

On the next day, Smith, with Dr. Roland M. Harper, botanist to the Geological Survey, was at Oyster Bay on the banks of "a deep stream, black water with saw grass and sagittauria border on opposite bank and here was some of the peat. Mr. Harper got out into this water a foot deep and probed down several feet easily through peat or muck. Most of what he has taken is pretty sandy and will probably be of no value—but no doubt some of these bordering marshes on fresh water creeks etc near tide water will yield some kind of peat possibly of value."

On June 12 while they waited for their guide to finish his dinner, Smith and Harper "ate a few ginger snaps and similar delicacies from the store" before embarking in the guide's motor boat "down river and into Bay St. John [Oyster Bay]. Here also Harper got some samples of peat—pretty dirty ball of it. The wind was . . . fresh and we got . . . well dashed with the spray when the bow of the boat struck a good wave. We did not go far down into Bay St. John but returned to Oak in time to take a . . . good walk northward from Oak through the piny woods. All day the route has been over flat pondy land with the usual vegetation . . . Mr. Harper found in several of the ponds a vigorous growth of Stillingia aquation not before identified in Alabama."

On June 14, Smith took the five-hour motor launch trip from Oak to Mobile leaving at six in the morning. Truman was in Pascagoula, but Smith went out to his house to have supper with Pearl, then they met Truman's boat at 10:30 p.m. He stayed with the couple until Sunday midnight when he caught the train for Tuscaloosa.

The last week of June, Landon assured his father that: "There is no truth to the rumors that I am going to marry." He added that he intended to continue following the plan that his father had outlined in a letter for paying his debts within sixty days.

On July 20, Smith wrote Merrill, who was in Mobile visiting Truman and Pearl, that Jennie had decided not to attend Merrill Edward Pratt II's wedding coming up in September "as she says she has no suitable clothes for an occasion of that kind. Still I don't see why the looking on at a wedding should call for any very fine clothes unless she was expected to take part in some reception or public entertainment in Montgomery."

Later in July, he wrote Merrill about his trip by train from Woodward to Tuscaloosa: "When we got nearly to Cottondale a crazy woman that they were bringing to the asylum got up from her seat about the middle of the car and ran to the front door and jumped off the train going about forty MPH. Of course, everybody expected that she was killed. When the train backed up however she was sitting on some cross ties at the side of the track and did not seem to be much hurt except she had struck her head against something and was bleeding from the place on her forehead. She was able to walk and get on the train. When she started to run to the front door the man who was sitting with her on the seat tried to catch her but she outran him. I can't see how she escaped for where she jumped off there were a lot of old cross ties lying about. I don't think anyone could do it again without being killed or else all broken up."

Not until early August did Smith ship Julia's piano and other furnishings to Truman with detailed instructions for unpacking the piano and with the added note: "Your mama seems to dread going anywhere. She's been under the weather for three or four days."

On August 9, Smith attended the gala opening in Tuscaloosa of the Tuscaloosa Mineral division of the Louisville and Nashville Railroad where a silver spike, driven in by a local belle, symbolized the completion of the road. The L & N's superintendent made the speech accepting the rail line after which a special train, according to the local paper, "carried the crowds to the barbecue grounds in the [Bryce] hospital grove [where awaited] . . . 800 feet of tables heaped with every thing good to eat."

The following month, Dr. Hugo deVries, distinguished professor of botany at the University of Amsterdam, arrived in Tuscaloosa to meet with Dr. Smith and Dr. Harper. DeVries was searching for the rare evening primrose that reportedly grew in Baldwin County, Alabama. William Bartram, who first saw the plant in 1778 not far from the confluence of the Tombigbee and Alabama rivers, described it as "perhaps the most pompous and brilliant herbaceous plant yet known to exist" being seven or eight feet tall and bearing hundreds of yellow flowers about five inches in diameter. Bartram's discovery was overlooked until 1904 when the New York Botanical Garden librarian called attention to the botanist's account. More than one hundred thirty years after its initial discovery the primrose was re-found growing, in full bloom, in the original area. Dr. deVries's interest in the primrose grew out of his Mutation Theory of Organic Evolution, which held that "new species of individuals, both animals and plants, could arise by sudden mutations rather than by fluctuating variations." While visiting Dr. Smith, he made a speech in Smith Hall on plant changes particularly regarding plant life in Alabama.

On November 13, Smith undertook a trip to Covington County for the Soil Survey traveling for the first time by automobile: "On Wednesday morning . . . we hired an auto and drove . . . up the country to Rose Hill and a few miles beyond (near Andalusia and Opp) . . . As we turned back to Opp it began to rain and this put the roads in such a condition as to cause us to send the auto back from Opp and abandon the trip."

The following month, Jennie entertained the Daughters of the American Revolution at the Smith home. The *Tuscaloosa News* reported: "The taste of the hostess was expressed in the appropriate Christmas decorations of the room where holly was used and in the beautiful centerpiece of the table which was formed of a pyramid of winter fruits with clusters of grapes and crimson ribbons giving another bright touch. Exquisite Cluny lace mats were used upon which rested cut glass compotes filled with crystallized ginger, almonds and stuffed dates. A delicious salad course with scalloped oysters in timbales, with hot rolls, sandwiches and olives were served first and . . . coffee with fruit cake followed. Mrs. J. Virgil Jones poured the coffee from a prettily appointed table in one corner of the room." The *Tuscaloosa Times-Gazette* added: "The house was more than usually attractive with an artistic arrangement of many ferns which were used as a background for . . . vases of pink carnations noted on every side of the livingroom."

Also in December, members attending the annual meeting of the Geological Society of America chose Dr. Smith as president. He delivered a paper, "Some Pioneers of Southern Geology, and their Contributions to Geological Literature," with mention, among others, of Michael Tuomey, E. W. Hilgard, and Daniel W. Langton.

1913

DEKALB, JEFFERSON, RUSSELL, AND TALLADEGA COUNTIES

Congratulations showered on Smith from every direction in January, for the honor to himself and to Alabama. Hopson O. Murfee, superintendent of Marion Military Institute, added: "My mother read with interest the article in yesterday's *Age Herald*, and spoke of my father's esteem and affection for you and his great appreciation of your life and work." Speaking of Jennie he wrote: "We have grateful hearts for her goodness to our children during their sickness in Tuscaloosa three years ago."

On January 6, Jennie herself wrote Pearl a news filled letter covering the Christmas holidays (she named her many guests); Merrill (he "seemed to have a very gay and happy time"); Papa (he had a fine time at his meeting—they had a big storm the day he arrived in New York—"the wind blew in four different directions by his count, . . . he had to leave off his glasses and hang on to his hat"), and Amelia Gayle Gorgas (she fell sick with grippe after Christmas and died in January). Smith served as honorary pallbearer for his family's old and dear friend. Dr. William Crawford Gorgas's duties in Panama made it impossible for him to attend his mother's funeral.

A variety of worries plagued Smith in February, March, and April. Truman had an eye problem; Smith wanted him to go to Montgomery and consult Dr. Charles Thigpen, "the best man in the country." For three days in Birmingham, Dr. Smith sat for Miss Edna Smith to paint his portrait for the Alabama Department of Archives. The Gorgas sisters (Miss Maria, Miss Jessie, and Miss Mary) expressed to Smith their desire for a special monument for their mother's grave in Evergreen Cemetery. Smith asked Truman and Pearl if Miss Jessie could stay with them when she came to Mobile to select the marker. She had no friends there, and "she would be pretty lonesome all by herself . . . The Gorgas family has been very close to us all these years as you know." And Merrill was failing calculus at the university. "I believe his main trouble is that he devotes too much of his time to society" was Smith's verdict.

Not only Merrill, but also all the students went wild over dancing. The university, however, frowned upon their favorites—the tango, the peanut stroll, the Gaby glide, the march of the dead, the turkey trot, the grizzly bear, the kangaroo lope, the bunny hug, the walking George, the Palm Beach dip, the kitchen sink, and the Tuscaloosa tussle—and sought to bring back the two step and the waltz.

The *Birmingham News* noted (April 20, 1913) that Dr. Smith, after thirty years of waiting, had finally come into possession of a curbstone from First Avenue in Birmingham. The block of sandstone, six feet by two feet by six inches thick, weighed over a thousand pounds. It showed on both sides "well-defined ripples or wave-marks, which were made by the waters of a Paleozoic sea years ago." Smith had noticed the stone, brought from the mountains of Red Gap near Gate City, soon after it was laid during the early days of Birmingham. "When I have had it thoroughly washed and scrubbed," Smith said, "it will present a much better appearance and will make an interesting addition to the museum."

At the end of April, the eye doctor ordered a month's rest in the mountains for Truman. Smith offered to schedule a wagon trip to Talladega, Clay, and other elevated parts of the state, but he could not leave before June. However, Merrill could not go that early, and he would be dreadfully disappointed because for years he had longed to make another wagon trip, but his father instead had insisted he spend his summers with a calculus tutor.

The Smith household overflowed with company during commencement in May. (The *Montgomery Advertiser* announced the largest enrollment in university history: 1,167.) In the confusion, Smith lost the pants and coat to his dress suit and was left with only the vest. He was sure one of the guests had packed them in his luggage by mistake, but the missing garments turned up sometime in July stored in a washstand drawer "in the room in which I slept during Commencement," Smith wrote Truman. His other news included a trip to Birmingham to have a new set of false teeth, "dental plates," made. "They do not yet fit well but will have to be ground off in several places."

At the end of July, Truman's eye continued bothering him, and he was, as usual, considering a job change. Smith recommended that he stay where employment was certain even if promotions came slowly. He also suggested that Truman stop the constant eye examinations, as the eye showed no change and he had become fixated on the problem. "Dr. Gorgas tells us that he kept sparing his eye in every way for a year or so, but after that paid no attention to the matter and his condition has been improving all along," Smith wrote. An additional worry was Truman's continued use of belladonna which, Smith feared, might be injurious. As an after thought, he wrote that his dental trouble had not been solved to his satisfaction.

In August, Smith was called to Gadsden as witness in a trial. He made the most of it by taking a side trip to Ft. Payne, traveling by automobile. On the return trip, "another auto came in sight and would have crashed into us but for the fact that the driver of the other machine ran off over a low wall into the bushes on the river bank. His machine had no brake control so that he could not stop his machine and if we had come together probably both machines would have been wrecked and gone over the abutment at the end of the bridge."

Meantime Jennie wrote Pearl about the Gorgas sisters leaving town, but they planned to "return in Oct. and live one more year in their old home. I am so glad as I should have missed them more than I could say."

The trial that Smith had come to Gadsden for was deferred. He seized the opportunity to motor with friends to the onyx cave at Kymulga. "The onyx is in stalactites and stalagmites lining roof, sides and floor of the cave. The stalactites are in conical pendents—corrugated sheets etc. Some pendents are many feet in section thus being in position to yield slabs etc. of any reasonable size. Stalagmite pillars also 4-6 feet more or less in diameter are in places built up towards the stalactites sometimes meeting them. The floor of the cavern below the coating of mud is also of stalagmite origin of thickness unknown, since no borings have been made into it. Pieces knocked off from various parts of the cave show that the prevailing material is light colored, creamy white or translucent with occasional red streaks. Other varieties show various shades of red and yellow—often sharply defined against each other, onyx fashion. The part of the cave explored by me was several hundred yards—following the windings.

"Several branch caverns are known but I did not enter any of them. The quantity in the main cave seems to be ample for supplying present demand in U.S. for onyx for many years to come but it would be well suited for all kinds of interior decorative work and so might find a much larger market than onyx seems to command at the present time. Samples of the material have been cut and are exceedingly beautiful."

On August 10, Smith wrote a postscript to his denture problem to Truman: ""My only trouble with the store teeth which I have . . . [they] are perfectly useless for chewing purposes though perhaps all right for appearance." In August, Smith was "laid off with malaria." On September 4, he wrote Truman, "I hope I have got over the malaria for good, but you never can tell. I am taking quinine as well as chill tonic."

In another September letter, he wrote: "We have just received a car load of things constituting Mr. Aldrich's latest donation to us. Ten cases, thirty or more drawers each, filled with his shell collection, and about forty boxes of books. This is a grand gift. I have also from Rev. Mr. Wheeler, now in Arkansas, who was one of our house-boat party two or three years ago, received a gift of most beautiful minerals. Besides we have just acquired by exchange a great lot of other desirable things so that the museum is forging right ahead."

In a follow-up letter, he assured Truman: "[T. H.Aldrich's] donation puts us in the very first rank in Conchology and at the very top in the matter of the land and fresh water shells of Ala., of which I suppose we have every species that has yet been described along with a lot of new species collected this summer by Mr. Herbert Smith [curator of the museum]. Mr. Aldrich [now postmaster at Birmingham] spent a couple of days here last week helping to unpack and place the drawers in their proper cases. These drawer cases, nine in number, occupy the spaces between the windows on the gallery on the west side of the bldg, Mr. Aldrich having given us the cases along with the shells."

The *Mobile Register*, calling it the "richest private conchological cabinet in the United States," enumerated "seven fine private collections" which Aldrich had acquired and included in the gift in addition to his own collections:

> The collection of Mauritian shells made by Col. Nicholas Pike, United States consul at Maritius
>
> The collection made by J. M. Jones in Bermuda and Nova Scotia
>
> The large collections made by the late Mr. Henry Hemphill in Florida and California
>
> The Parks collection of about 5,500 listed species, many of them of great rarity
>
> The collection of W. W. Calkins, especially rich in American fresh water forms
>
> The duplicate collection of Foote, about 2,000 species
>
> About 1500 species from the Philippine Islands, purchased from Walter F. Webb

The same newspaper reported that among the cases of books Truman Aldrich donated were "many . . . rare and costly works on conchology and other branches of science . . . For many years Mr. Aldrich has been a generous friend of the University Museum. Some time since he gave his entire set of duplicates, some 200,000 specimens of at least 4,000 species, and he has also donated a series of tertiary invertebrate fossils so rich that it placed that department of the museum in the first rank. But this latest gift is the noblest of all, and it comes to the university primarily as a tribute to Dr. Eugene A. Smith, for many years a close personal friend of the donor."

In Russell County near Pittsview on September 10, Smith was again traveling by automobile and having a difficult time because of rain and deep sand. "We . . . started to Uchee passing Sand Fort—but a little beyond that the pulling was too much for the car and we had on the low speed all the time with the result of getting nearly everything red hot and water boiling and requiring to be renewed every ten to fifteen minutes. Mr. Bell, who knows the way, said the sand got continually deeper and he did not think we could pull through it. So we turned back by Sand Fort and took a cut across to Seale-Hurtsboro Road—which though worked and graded was surfaced with clay and we fared no better as the car could barely pull through it. We intended to turn up the Uchee road when about four miles from Hurtsboro but concluded we could not make it as we were traveling only about three miles an hour."

Jennie visited Truman and Pearl in October. While she was there Smith wrote Truman: "The [university] boys got off this morning to B'ham, five [railroad] cars full. They got aboard here at the univ. which is a great improvement over the old way by the AGS [depot]. I don't know how they will come out with Georgia as it has a fine team. I think 300 must have gone . . . [as] there were only 65 at dinner at the mess hall where the usual number is over 250. Merrill . . . did not go. Merrill and Owen . . . are unloading a car of coal at the powerhouse for which they get $3. It will take them eight hours solid shoveling. Merrill looked like a coal miner at dinner time. I think they will earn their money."

On October 23, Smith wrote Truman: "I suppose you have seen by this time a notice of . . . [Aldrich's] marriage either yesterday or the day before. We had heard that it was going to be though he had not notified any of us." The first Mrs. Aldrich, about whom little is known other than that she was an

Episcopalian, had died the year before. Now Mr. Aldrich married a Birmingham widow, Helen Harris, the mother of four daughters.

During Dr. Smith's fourth decade as Alabama state geologist, he coped with a great deal of stress in his personal life. In his profession, however, he received more and more honors and recognition and many valuable gifts came to the museum because of him. The allocation of a salary eased his financial situation somewhat and probably pleased Jennie, not only for the money but, more importantly, for the validation of her husband's devotion to his work. Of the twenty-one publications that Smith was responsible for, the subjects of cement and water resources dominated. He also wrote of the Cretaceous-Eocene contact of the Tombigbee River, the post–Eocene and other formations of the Gulf region, the iron ore industry, Smith Hall, and memorials to Henry McCalley and Daniel W. Langton. Now at the beginning of Dr. Smith's fifth decade, he and the curator of the university museum, Herbert H. Smith, set in motion an ambitious project involving the schools of Alabama.

5

The Fifth Decade, 1914–27

1914

GREENE, MOBILE, AND TUSCALOOSA COUNTIES

Between 1910 and 1914, Eugene Allen Smith and the curator of the museum, Herbert H. Smith, got underway an elaborate "scheme" to place free of charge in the high schools of Alabama what amounted to branch museums. The purpose of the project was to "arouse the interest of the rising generation in Natural History the educational value of which is very little appreciated by school boards as a rule," the initial announcement said. It was intended that the gift "be continuous and of indefinite duration."

The first donations, which were distributed in 1910, consisted of twenty-five sets of one hundred fifty species of shells chosen from the duplicates that had been donated to the museum. The sets were "provided with correct labels, paper trays," and meticulous directions for the care of the specimens. The first installment was of marine shells "not because we consider them more important," the announcement said, "but because our duplicate stocks were large and immediately available . . . We hope the second set issued will be Alabama fossils. We plan to give fossils enough to illustrate each great formation as well as a good series of rocks, minerals and ores."

In 1914, Herbert H. Smith evaluated the project thus far: "Most of the recipients made good use of their gifts. Curator has received scores of letters from school pupils in every part of Ala. Many of them childish but many showing keen powers of observation and all overflowing with enthusiasm. Requests for collections came from other schools. In order to aid these we offered to help schools in building up their own cabinets. This scheme has worked well. Pupils collect fossils, shells or other objects, send to us for naming; they are returned properly labeled with additions from our stock; should be supplemented by visits to schools, excursions with the curator, illustrated lectures, etc. We are hampered by the total lack of funds. All we can do is to visit some of the schools when our collecting trips bring us to their vicinity. School parties have been taken out on excursions with excellent results. In 1912 a camp on the Cahaba River with boys from Montgomery, Dadeville, and Tuscaloosa spent a week with the Curator. April 1914 Eugene A. Smith formally organized at Tuscaloosa an Association of school pupils who love nature, about a 100 members."

In the early months of 1914, Landon surprised his parents by bringing Jerome Chapman of Tupelo home to marry. Jerome, just out of high school at age sixteen, was the daughter of the head of the Tupelo Military Institute. Merrill took the part of groomsman to the thirty-three year old Landon. Immediately after the ceremony, the couple left Tuscaloosa for Tupelo where they planned to live. The newspaper account noted: "The plans for a brilliant reception were necessarily changed on account of the groom's illness."

At the 1914 commencement, President George H. Denny presented diplomas to veterans whose education at the University of Alabama had been terminated by the Civil War. As Smith had been actively involved in training the cadets, he probably knew all of those honored. He saved this newspaper account of the ceremony: "Tears came to the eyes and throats were choked with emotion as a thin gray line of soldiers of the Confederacy came forward in Morgan Hall . . . to receive their diplomas . . . deferred a half a century . . . 150 degrees [were] conferred. Only 25 of the old students were present in person, but about 75 diplomas were given to the representatives of the honored veterans. The remaining will be mailed. The ceremonies were centered around the unveiling of the boulder erected by UDC."

The following month, Smith attended a called meeting of the Alabama Highway Commission in Montgomery. His sudden concentration on gravel sources may have grown out of this meeting as more and more automobiles were appearing on Alabama's "execrable" roads. He continued to Mobile for a visit with Truman, spending a day with him "on the boat *Dauphin* down the Bay." On the way back to Tuscaloosa, he stopped off at Glendon to look over the gravel pits of the Mobile Gravel and Material Co. The pits produced a clay gravel "very suitable for use for road surfacing. The daily production maximum is 5 or 6 fifty-ton car loads a day. A shortage of cars prevents the company from making the max. production."

On March 20, Smith went by train "to Shiras (Ruhama Church) to look at gravel beds in abundance high above [the Black Warrior] river." Still on the track of gravel, Smith next went by train to Boligee, then, traveling with others by automobile, he "rode down through Forkland to the Indian Mound." Smith arranged to send a map of Greene County to Mr. Thornton of Thornhill who agreed to "put in all the gravel capped hills he knows." They next visited a gravel pit on the Hohenlinden farm, just off the Eutaw road. The auto had difficulty traversing the "deep ruts worn by gravel hauling—we ran against a stump in the middle of the road which caused a bending of front axle and breaking of front spring cuff, besides giving us all a thorough shake up. Fortunately we were going uphill and slowly." Smith and another member of the party left the disabled vehicle and walked four miles to the nearest railroad station, arriving there just in time to catch the train.

Of his visit to Rosemount, the former home of Williamson A. Glover, Smith wrote: "The house is a magnificent example of an ante bellum residence—now fallen into utter decay—a depressing spectacle. A few neighbors were having a picnic at this old place, as we were." Built in 1832, the twenty-room Rosemount was crowned by a cupola with a colonnaded walkway. Thornhill, not far away, had fared better. Smith called it "a splendid type of ante bellum mansion . . . kept up and . . . in good condition and well worth seeing."

1915

CLAY AND CLINTON COUNTIES

Graphite, as it had many times in the past, drew Smith to Ashland in June. Dr. C. S. Northen met him at the station with his auto and took him to his Alabama graphite works. "The mill and plant generally is efficient, having been worked out by Mr. Griesemer who, they say, is the only man who has a process that will clean the graphite and save the greater part of it. From the Alabama works we went over to the Quenelda establishment. Here again a very well constructed plant and with the Griesemer method. The present Quenelda mill is about a mile or less from the original Quenelda now abandoned. The quarry is half a mile away with dinky train. The graphite bearing rocks here stand practically vertical and the width of graphite rock from 80–100 feet. The old Ashland graphite plant between these and town is now idle. The original . . . graphite plant also abandoned—all the buildings destroyed by fire.

"Dr. Northen promises to forward along with the ore specimens collected a series of samples of the different grades of graphite from Alabama and Quenelda works. He will also give me the exact location of all the works that have been in operation in Clay County together with a history of their ownership, etc."

In August, Smith returned to Ashland: "Mr. Mattison came with his Ford and we went first out to the quarry and graphite plant of Jennings and Mattison Clay Co[unty] Graphite Co. . . . In both these the graphite is in two veins about one thousand feet apart and one hundred feet thick. . . . The present mill according to Mr. Mattison produces about 20, 150 lb bags of flake graphite daily. . . . We went down to Verbena and out to the Flaketown Graphite quarry and mill. . . . The process here is dry—no water used. The ore is crushed between rollers—so as not to break or injure the flake—then subjected to air blast which separates most of the sandy material, then treated with electric plates 11,000 volts by which the graphite is still further removed from sand. The graphite is more quickly electrified than the mica and this is the principle involved. The tailings from the first treatment still contain some mica and is treated again with electricity and additional 2000 volts. The graphite produced is only of one grade." The end product was widely used in the manufacturing of paints and lubricants and in the making of pencils.

The life-size oil portrait of Dr. Smith, painted by Edna Kenderdine Smith, was presented to the Alabama Department of Archives and History in 1915. Noted for her portraits of well-known Alabamians, Miss Smith (no relation to Dr. Eugene Smith) came to Birmingham from Illinois. She graduated from high school in Birmingham, studied in France several years, and opened her studio in Birmingham in 1902.

1916–20

BIBB, CHEROKEE, CLARKE, CLAY, COLBERT, JEFFERSON, MADISON, MONTGOMERY, SHELBY, ST. CLAIR, SUMTER, TUSCALOOSA, AND WILCOX COUNTIES

In Montgomery again on September 8, 1916, Smith wrote: "At 3:00 p.m. meeting in Governor's office to organize the Alabama Board of Trade and later a meeting of the State Highway Commission to consider some contracts, and the letter of the U.S. Secretary of Agriculture, in regard to government appropriation for good roads."

October found him in Monroe County: "At Mr. J. M. Kearley's near head of Curiosity Branch, I saw some very interesting specimens, among which there were two of a spherical ball in hemispherical bowl—the ball 12 inches or more in diameter and the bowl nearly 24" diameter. The ball is loose from the bowl. I suppose the weight of ball and bowl will be over one hundred lbs. The bowl is broken into sections by cracks and is one half of a spherical concentric envelope of the central ball." He sketched it in his journal.

The *Montgomery Advertiser* reported January 2, 1917, that Dr. Smith was in Montgomery for a meeting of the State Highway Commission. He also called

Dr. Smith's photograph of the "spherical ball in [a] hemispherical bowl."

on state officials to lend his influence against ratification of the proposed Prohibition amendment to the U.S. Constitution. The *Advertiser* wrote: "He's always been an important factor in Alabama politics though rarely partisan. He is an old Confederate soldier though and in this fight sees war on the principle of state's rights for which he struggled." The story continued: "Dr. Smith is perhaps the most popular man ever connected with the University and is beloved by every student that ever attended the school. When he walks down the streets of Montgomery he is stopped and cordially greeted every few yards by some former student of the University . . . He has a wonderful memory for faces and names and it is a very rare thing that he fails to call a man by his name right off the bat."

In June of 1917, Smith and his party drove a "zig-zag route" over Jefferson and Shelby counties. They visited Brierfield where Smith "saw the foundations of the old ordnance works of the Confederate Government. There are very few houses left of the old settlement; the chief remnant of the past being a brick store."

Smith called on his old friends Dr. Thomas Palmer, third president of the Girls' College in Montevallo, and his wife. Dr. Palmer formerly taught mathematics at the university. He showed Smith through Bloch Hall, the new science building. Smith noted the building was not only fire proof, but well-planned and well-equipped as well.

On this trip Smith wrote: "The ore pits at Shelby are something enormous—in places over one hundred feet deep and half a square mile in extent." At Ragland they "drove out to the quarry which is in the side of a pretty good sized hill . . . The limestone is hauled by aerial cable to the cement plant—about three-fourths mile. At the plant I met Mr. Ernest the chemist, with whom I had had correspondence. The laboratory is well equipped for cement needs." On June 9, they "came through Village Springs which has been practically wiped out by a tornado. Only one or two houses are left, and the whole place is a tangled mass of broken timber and plastering."

Later in June, Smith visited the pyrite works in Pyriton. He found a couple of mines being reactivated with great promise. With Mr. Bugg and Mr. Edwards of the A B & A Railroad and the two owners from Birmingham, he inspected the National Pyrite and Copper Company. "The engineer in charge . . . has just cleaned out the slope and will be ready to get out some ore in a week or ten days. He thinks he can get out fifty tons a day . . . [He] will not put up a concentrator until he finds other commercial lenses of the ore on his holdings. Took dinner, along with the others, with Mr. Bugg on his private car. This trip was made with the purpose of meeting Mr. Fuller Callaway of LaGrange, Georgia who has been asked by the Bureau of Mines to get up notes on the pyrite possibilities of Georgia, Alabama, and perhaps other southern states."

On July 23, 1917, Smith arrived in Piedmont. Here he met T. C. Cowan "who has a theory about oil and gas in this section. As well as I can make out he depends on a line of 'gas blowouts' which encircle an area under which he is confident that there is a great body of oil. The 'blowouts' follow Knox Dolomite chert ridges and to me appear to be only lime sinks . . . We drove northward to a bauxite mine belonging to the National Bauxite Company . . . The quarry is some fifty feet deep and about as wide on the surface. The bauxite . . . would have had no market at all a few years ago, though some of the ore is white and pure like the best of the Rock Run ore. They haul it six and a half miles to the RR at Piedmont."

On June 27, 1918, Smith wrote: "Stopped off at Cuba . . . and took an auto to see the clay beds near Kewanee that supply the Cuba pottery. About four to five miles NW of Cuba this clay shows in a number of places by the roadside."

A month later, he caught the train to Rock Run for a three-day trip: "Left Birmingham by 2:30 Seaboard for Piedmont where I was met by Mr. J. M. Garvin's car and brought to Rock Run. On the 26th Mr. Garvin, Miss Houseal and I made auto trip through Center, across the river by Bradford's ferry and up the valley parallel with the RR from Round Mountain to Jamestown [on Little River Canyon]. Here is a lead deposit which a company is preparing to work. A shaft has been sunk and boilers, engines are being set and other preparations are in progress. The galena is very sparingly disseminated through Trenton Limestone, so sparingly in all that I saw, as to make it seem to me impossible to extract with profit . . . I was again struck by the number of great waterworn boulders of the quartzite about Rock Run and along the way to Piedmont.

"In Birmingham I had two hours between trains and went out to Highlands to see Jerome [Landon's wife] and baby and came home by the 4:10 AGS train." Landon, with two degrees from the University of Alabama, one in civil engineering and one in mechanical engineering, served with the American military units in France. After the Armistice in November 1918, he remained in France to help rebuild the roads. He attained the rank of captain and received an honorable discharge in 1919.

A heavy blow to Smith fell on March 22, 1919, when Herbert H. Smith, curator of the museum, was killed at 7:30 a.m. by a freight train while walking to work at the university. The *Tuscaloosa News* reported: "Dr. E. A. Smith spoke most feelingly of the loss the University museum experienced with the death of Dr. H. H. Smith. Dr. E. A. Smith said it will be a hard matter to get a man to take his place as he was conversant with the immense quantities of specimens now on hand to be distributed half and half with institutions and individuals who put up the money to gather them." E. A. Smith announced that Dr. H. H. Smith's wife, who "knows more about this part of the work than any other living person . . . has consented . . . to remain here for a few months to help make the distribution and to attend to other matters with which she is conversant." Mrs. Smith worked closely with her husband; she had been his ears for he was very deaf. She traveled with him over much of the world collecting and studying, and worked by his side at the Carnegie Museum and later at the University of Alabama museum. The unfortunate morning of March 22, he had walked to work without her and had not heard the train coming. Eventually the Reverend Mr. H. E. Wheeler, a noted collector who wrote up the 1908 river trip, assumed leadership of the museum.

The Association of American State Geologists arrived in Birmingham August 31, 1919, "the first meeting of this distinguished body of scientists to be held in Alabama," The *Montgomery Journal* reported, "and Dr. Smith is very anxious that they be given an opportunity to see Alabama's wonderful possibilities along lines in which they are deeply interested." Between meetings held at the Tutwiler Hotel, the group visited local points of interest. The second day they divided into two groups one of which remained in Birmingham to visit Tennessee Coal, Iron and Railway Company (where they were served an old fashioned barbecue dinner at noon) and the mines of Republic Steel and Iron Company. The other group went to "the graphite fields in and around Clay County," according to the *Birmingham Ledger.* The two groups reunited in Sylacauga to tour the marble quarries and to continue northward stopping off at various places to be dined and entertained on the way to Muscle Shoals.

In the Sheffield area, they inspected "the government's nitrite industries, blast furnaces and the Wilson dam enterprise at Muscle Shoals," according to the *Birmingham Age-Herald*. "A portion of the cofferdam at Dam No. 2 has been pumped dry, showing the bed of the Tennessee river, and this will enable the geologists to view the dam construction work at a very interesting stage."

The *Huntsville Times* reported: "Huntsville has never had the honor of being visited by a more intellectual, distinguished and useful body of men." The big spring caught the interest of the group, the *Huntsville Telegram* wrote. "Quite a while was spent at the spring, the visitors soon scaling the cliff and securing specimens for their collections. They expressed wonder at the great volume of such clear water." They visited thoroughbred horse farms, textile mills, and many other places in the Huntsville vicinity, the *Telegram* continued. "Several of the party were taken so with the mountain trip that they asked to be left there to continue their examinations until the rest of the crowd finished their rounds. This was done and they were sent for later in the afternoon. Not only did the research party on the mountain disclose traces of limestone that they did not expect to find here, but some Shittim wood trees that grow only on this mountain and in the Holy Land were found."

The geologists returned to Birmingham to take the train for Tuscaloosa on September 6. They stopped off at Searles and Kellerman mines on the way to Holt where they visited the furnace and by-products plants then took the trolley to the university for further meetings and lunch at Smith Hall. Jennie was one of several faculty wives providing the lunch. In the afternoon, they were driven about the city and went to the riverfront "to see the locks and the quarry and the many coal pebbles and fossil plants," the *Tuscaloosa News* wrote. Most of the visitors departed on the afternoon trains.

To assist soldiers returning from the World War who wanted a farm under the land grant act, the Geological Survey published a report on what Alabama could offer. The *Tuscaloosa Ledger* wrote that the report covered "the size and value of farms, the character of the soil, livestock, the commonest farm products, and natural plant life . . . for every county and section in the district covered. Knowing that the soldier-farmer would like to know also something of the people who will be his neighbors the report gives information concerning proportion of different races, the number and distribution of foreign-born, and the ancestry of the people . . . One of the interesting parts of the report will be that giving the educational and religious conditions of the people. The percentage of illiteracy and the predominating religions are given for each county." Dr. Roland M. Harper compiled the work, and Dr. Eugene Allen Smith "is in Montgomery now endeavoring to secure the appropriation so that the report may be ready as soon as possible for its intended benefits to ex-soldiers and others."

Landon had returned from the war and in October 1919 arrived in Montgomery to begin work as engineer with the State Highway Department. "[His] addition to the Highway Department is considered a good move on the part of highway engineer W. S. Keller," the *Advertiser* reported. Jerome and the baby did not come with him.

Smith's salary as state geologist was increased in 1919 to four thousand dollars annually.

Early in 1920, the *Montgomery Advertiser* announced two exceptional gifts to the University of Alabama Museum in honor of Dr. Eugene Allen Smith. Dr. J. W. Spencer of Washington, D.C., the former Georgia state geologist and a well-known botanist, donated a collection of fossils and natural history specimens, and Dr. Truman H. Aldrich of New York, former Birmingham postmaster and Alabama coalmine operator, gave a collection of tertiary fossils found in Florida. Both men, the paper pointed out, had previously given other collections to the museum because of their regard for Dr. Smith.

In July 1920, Smith, working with Alabama's U.S. Senator Oscar Underwood, established a metallurgical research experiment station at the University of Alabama under the auspices of the United States Bureau of Mines. The *Birmingham Age-Herald* reported: "A superintendent and six metallurgical chemists will be assigned to the station to cooperate with specialists of the University. It will involve an expenditure of $25,000 annually, the library and equipment being furnished by the University . . . The work of the station will be to make researches and experiments in non-metals. Some of the important non-metals of which Alabama has a large share are graphite, cement materials, bauxite, phosphates, clays and feldspar . . . Results of researches will be made public; as a result large industrial concerns dealing in non-metallic

View from Smith Hall in 1921 shows Tuomey Hall, named for Alabama's first state geologist, on the left. Morgan Hall, with an exterior matching Smith Hall, is in the center. Clark Hall and Garland Hall, named for Jennie Smith's father, are on the right.

substances will have full advantage of immediate analysis and improvements made in chemical methods."

November 3, 1920, Smith went by train to Selma where he spent the day socializing and catching up on the local news before continuing to Thomasville. He noted in his journal: "The weather today is all that one could wish. The papers announce the election of Harding to the Presidency." The next day he wrote: "[This] morning we went in a Ford out in the direction of Sunny South . . . along an abandoned road . . . Returned to Thomasville about noon but did not wait for dinner but made a trip westward with Mr. Roy Kimbrough in his big car . . . Came back to Bashi Store then southward to Tallahatta Springs . . . but we could not find the spring because of rank growth of weeds and underbrush . . . The place is utterly neglected. The old Bedsole place, a two story frame bldg is here, not perhaps one hundred yards from the spring."

1921–27

Autauga, Clarke, Dallas, Jefferson, Marengo, and Tuscaloosa Counties

At the beginning of Smith's eightieth year, he met in Birmingham with George H. Clark, T. H. Aldrich, E. E. Ellis, and others to consider the most important and pressing industrial problems of the area. The aim of the discussion, Smith wrote, was to "become acquainted with those problems from [the] point of view of the heads of the great plants, and thus be in position to take these up in the Experiment Station work in order of their importance." The group began a tour at Selma, putting up at the Albert Hotel. "On Friday 20th we came in several cars out to Plantersville and spent most of the day in that vicinity and over about Vine Hill looking at the outcrops of manganiferous sandstone about which some interest has broken out." They drove in the vicinity of Quarles and the Plantersville High School, then "eastward across the creek and then down on the left side of same about to the old Lapsley place. The Selma people, Crissman and others, have surveyed all this territory and divided it up into moderate sized lots. This has been done some years ago but there seem to be no buyers or settlers. At the house of two bachelor brothers Harville, we got a very good dinner for all nine of the party, having telephoned in advance from Plantersville."

Smith traveled by train to Whatley on August 9, 1921, to meet Mr. DeKalb who came up from Mobile. At Pine Hill, Smith saw briefly Walter B. Jones who would become assistant state geologist in 1924. On to Coffeeville where Smith and DeKalb joined a large crowd attending a Masonic meeting. After several speakers, all of whom Smith knew, the two men partook of the "big basket dinner where all kinds of good things were set out on rough tables under the trees," he wrote.

After dinner Smith and DeKalb went "down to the Bluff where two cuts have been made . . . exposing about fifty feet of the . . . Buhrstone clay, or flint clay. The material . . . [has] the qualities of Fullers earth throughout." DeKalb spent August 11 at White Bluff taking samples, while Smith and Ed Cowan went by boat "past the mouth of Witch Creek" looking at the strata of the Hatchetigbee sands. Smith spotted a large "symmetrical and beautifully marked" concretion typical of the bluff that Mr. Cowan promised to retrieve and ship to him.

Their skiff "passed the 'Merry Mac' of Tuscaloosa, a sixty foot yacht owned by Faucettte and others . . . I could not help wishing that I might make a trip in it some time," Smith wrote. At another cut, they "found a horizon in which there were numerous vestiges of plants in sandstone or quartzite . . . Someone at our camp gave me a fine piece of silicified palm or cycad trunk very perfectly marked, probably from the . . . bed above mentioned."

The two men continued up river on August 12 to examine "another high bluff to the west of the cuts and got some beautiful tufts of fibrous gypsum from the joint openings in the siliceous clay like rock. This is like the finest asbestos but with fibres three-fourths to one inch long. We had the lunch brought out to the bluff . . . and came back to camp rather early, packed up and cleaned up everything and came on back to Grove Hill arriving about 7:30 . . . [That] evening we went out to Mr. Cowan's father's house to supper and met members of his family, including his sister who has taught in Northport this past year. Mr. Ed Cowan is not married and his father was not at home that day."

Jennie and Eugene celebrated their Golden Wedding anniversary in July 1922. A newspaper account mentioned: "Their eldest son Landon is with the state highway department in Montgomery and the other two, Truman and Merrill, are in engineering work in Mobile—a small grandson, Landon Jr. spends about half his time with Dr. and Mrs. Smith." Later that year, a reporter described the eighty-two year old Smith in these terms: "looks about fifty," "small in stature," "no sign of frailty," "kindly eyes with wrinkles that come from smiling often," "skin brown, not with sallow color of age but the healthy tan of the outdoors," "as interested in the human side of University life as he ever was," "still a member of the faculty Committee on Athletics," "still attends football and baseball games—whenever anything is doing on campus he's pretty certain to be there."

The following year, reminiscing on his eighty-third birthday, Smith told a reporter that the university had more faculty members now in October 1923 than there were students "back in the Seventies . . . It took us so long, nearly forty years, to bring the enrollment up from 55 to 300, and now in the past few years it has jumped up to 1800."

In 1923, Smith had a part in clearing an innocent man who had been imprisoned since 1915 when he was thirty-five years old. The case was even more complex than the Statesville poisoning. Tom Lewis, a black man from Sumter County, was sentenced to the state prison for murdering a wandering black minstrel. He was convicted on the testimony of a small black child who claimed that he saw Lewis and another man load the dead itinerant's body on a mule and haul it across a field and bury it near a cypress pond. Later, the child said, he saw them dig up the body, stuff it in an old sweet gum stump, and burn the stump. Bones were found in the stump and assumed to be the minstrel's. No motive came to light, but Lewis was convicted. However, later a man reported seeing the minstrel in a distant town performing in his usual manner, very much alive. The state geologist was then consulted. He stated, according to the *Montgomery Advertiser:* "To cremate a large body like that of a man, would require a degree of heat and a duration of it which could certainly kill a tree . . . and . . . this heat would consume a tree long before the body was cremated." He went on to say that to cremate a body required from twenty-five hundred to three thousand degrees Fahrenheit while a fire in a stump could hardly reach five hundred degrees Fahrenheit. Witnesses who had examined the tree stated the tree was still living when the bones were removed from it. Later a university anatomy professor identified the bones in question as those of a young calf. "It is likely," the newspaper stated, "that the board of pardons will recommend clemency for Lewis."

By unanimous vote in April 1924, Dr. Smith was made honorary dean of the newly organized Alabama Academy of Science, reported the *Tuscaloosa News:* "Dr. Smith related for the new organization the history of the former Alabama Industrial and Scientific Society."

Smith's writings and speeches were fewer during these years. The subject matter often had to do with mineral production and more and more with gas and oil. He wrote memorials for his old friends, E. Waldemar Hilgard and Robert Loughridge who used to send birthday gifts to the small Julia.

Bringing to a close fifty-four years of dedicated service to his state and his alma mater, Eugene Allen Smith died in 1927 just short of his eighty-seventh birthday. He was suddenly "stricken on August 29 while at the breakfast table with an internal malady and had to be taken to the hospital for an immediate operation," wrote Dr. Walter B. Jones in *Science.* "The operation was successful, but his vitality was too low to allow him to recover from the shock. He succumbed at 4:30 P.M., on September 7, and was buried the following afternoon."

The account given September 15, 1927, by the *Crimson-White*, the school newspaper, briefly summarized, read thusly: Dr. Eugene Allen Smith, oldest member of the university faculty, scientist of national repute, died after an illness of eight days following an operation for a strangulated hernia. Funeral services were held Thursday, September 8, at 4:30 o'clock at the Smith home. [Accounts in the Tuscaloosa paper said the funeral was in the chapel at Clarke Hall]. Benjaman Harrison, lay reader of Christ Episcopal Church of which Dr. Smith was a member, conducted services, he was interred in Evergreen Cemetery. Pallbearers were men Dr. Smith worked side by side with for the past several years in Smith Hall, at the university, and in fieldwork: Dr. George H. Denny, president of the University; R. M. Harper of the State Department of Forestry; Walter B. Jones, assistant state geologist; Dr. R. S. Hodges, chemist for the State Geological Survey; H. E. Wheeler, curator of Alabama Museum of Natural History; Dr. S. J. Lloyd, head of the Department of Chemistry. The account gave excerpts from tributes to Smith at the service, a biography, and a lengthy list of his accomplishments and honors.

According to Dr. Walter B. Jones, Smith's successor as state geologist: "As a result of . . . [Dr. Smith's] final request of the legislature, the annual appropriation was increased to $50,000, and his survey was assured of adequate support for the first time in its history. The final passage of this bill was recorded during the first day of his fatal illness, but he had full knowledge of its successful voyage through the law-making channels."

Experienced political negotiator that he was, Smith probably wished he had been in Montgomery to shepherd the bill through the channels as he had so often done. In any case, he would have been pleased to leave his successor in a more financially secure situation than he himself had ever been.

Jennie continued living in the house on the campus, until at age eighty-four, she died at 1:30 p.m. on December 15, 1930. She had been ill since the previous June when she had a paralytic stroke. Her sons—Landon and Merrill of Birmingham and Truman of Mobile—and her two grandchildren—Landon Jr. by Landon's first marriage, and Marguerite by his second marriage—were at her bedside. The funeral, conducted by the rector of Christ Episcopal Church,

was held in the Smith home. Burial in Evergreen Cemetery followed. The *Tuscaloosa News* reported that Mrs. Smith "was much beloved at the University by students and townspeople alike."

The most lavish estimates would not be excessive in summing up what Eugene Allen Smith accomplished (and endured) in the face of formidable obstacles. He left a solid legacy for Alabama's continuing economic development through the state's natural resources—soaring coal production from almost nothing to over twenty million tons annually, booming iron and steel industries, flourishing textile complexes and other water powered mills, cement plants, fertilizer manufacturing, to name a few. "Industrial plants emerged from the hillsides like mushrooms," Dr. Jones said. He believed that the "phenomenal development of the mineral industries of Alabama, which accompanied the progress of the work of the Survey, must have given . . . [Dr. Smith] a great deal of pleasure and satisfaction," but, modest as he was, Smith never in any way showed such pride.

Undated portrait of Dr. Smith by unknown photographer.

Smith's memory remained alive in the following years through printed and spoken tributes to him on important occasions, and, of course in the beautiful building bearing his name. The battered geology wagon, stored in Smith Hall's basement for years, turned up missing one day. Dr. Jones traced it to a local salvage dealer who had bought it for $15. He re-bought the wagon and returned it to Smith Hall. In 1975, he inaugurated a campaign to raise money for its restoration. With the aid of a $5,600 gift from Landon Jr., the wagon was brought to such a pristine state one would never suspect the many miles of bone-jolting roads it had traversed in every county of Alabama or the many injuries inflicted on it by uncooperative mules. It now occupies a place of honor in the museum along with other Smith memorabilia including an exquisite set of miniature tools that he carved when he was twelve years old. The Smith house and garden were razed in 1949 to make a site for Gallalee Hall, the physics building.

There are those who swear that Dr. Smith still wanders the corridors of Smith Hall especially late at night, turning on lights, opening and closing drawers, and whistling a faint tune. Instead, that could be Langton, returned to the work he loved, or faithful McCalley, but never Eugene Allen Smith. He would scoff at such a notion. Smith might be satisfied to know that he is so well remembered as geologist, mapmaker, botanist, photographer, educator, scholar, road improver, musician, drama and art appreciator, political savant, friend, family man, and all of the other things that he was to so very many people. Above all he was at heart a Wayfarer as he termed himself—he often began planning for the summer field trips in January of the year, impatient to take to the road again. He relished the masculine world of the outdoors, of the rough camp life, the company of his sons and the men who invested their lives in the hapless search for riches, the powerful industrialists who used his reports to build their empires, and the politicians carrying on their intrigues in Montgomery. Smith was a man of action. Anyone who ever knew him would be certain that even if he is not resting in peace beside Jennie in Evergreen Cemetery, he is not spending his time in aimless hauntings.

Bibliography

Magazines

Aldrich, Truman H. "Doctor Smith, An Appreciation," *Arrow Points*. Vol. 7, No. 5.

California Alumni Weekly. University of California, Berkeley. April 18, 1914.

Center, Clark E. Jr. "The Burning of the University of Alabama." *Alabama Heritage*. Spring 1990, No. 16.

Dean, Lewis S. "Alabama Gold: Golden Harvest of the Piedmont." *Alabama Heritage.* Summer 1991.

DeJarnette, Don. "A Day in the Field with Eugene Allen Smith." *Geo-Petro Notes.*Vol. 1 No. 2, 6-7. November 1975.

_________. "Early Days of the State Geological Survey." *Arrow Points*. Vol. 7, No. 5.

Hall, John C., and Frances Osborn Robb. "Eugene Allen Smith and the Geological Survey of Alabama." *Alabama Heritage*. Summer 1994.

Jones, Douglas E. "Doctor Koch and his 'Immense Antediluvian Monsters." *Alabama Heritage*. Spring 1989.

Jones, Walter B. "Eugene Allen Smith." *Engineering and Mining Journal.* December 17, 1927.

_______. "Eugene Allen Smith." *Science*. January 6, 1928.

Sartwell, Alex. "Tuomey Pioneered Alabama Mineral Studies, 1848-57." *Geo-Petro Notes*. Vol. 1 No. 2. November 1975.

Sellers, Dodd. "Special Order Studebaker – 1873 Style." *Turning Wheels*. Vol. 15 #1, January 1983.

Smith, Eugene Allen. "Biographical Sketch of Henry McCalley." *American Geologist* 35.

_________. *The Age of Steel*. January 3, 1891.

Ward, Robert David and William Warren Rogers. "Punishment Seven Times More: The Convict Lease System in Alabama." *Alabama Heritage*. Spring 1989.

Wheeler, H. E. "A Geological Vacation." *Alabama Christian Advocate.* April 12, 1908.

Willard, Jennifer. "Giuseppe Moretti." *Alabama Heritage*. Spring 1991.

Newspapers

Birmingham Age-Herald. Birmingham, AL April 25, 1894.

________. May 26, 1897.

________. June 13, 1906.

________. October 21, 1908.

________. August 21, 1909.

________. May 23, 1910.

________. August 15, 1912.

________. December 15, 1912.

________. October 5, 1913.
________. January 22, 1915.
________. September 19, 1915.
________. August 31, 1919.
________. September 1, 1919.
________. September 2, 1919.
________. September 3, 1919.
________. September 5, 1919.
________. July 15, 1920.
________. August 19, 1920.
________. July 10, 1922.
________. March 13, 1924.
________. June 4,1924.
Birmingham Ledger. Birmingham, AL. May 26, 1906.
________. September 29, 1913.
________. January 21, 1915.
________. January 22, 1915.
________. September 1, 1919.
________. September 2, 1919.
Birmingham News. Birmingham, AL. June 8, 1895.
________. February 9, 1910
________. April 20, 1913.
________. July 9, 1922.
________. January 20, 1924.
Chicago Herald. Chicago, IL. March 21, 1892.
Chicago Tribune. Editorial Tribute. Chicago, IL. July 6, 1883.
Crimson-White. University, AL. October 21, 1902.
________. September 15, 1927.
Demopolis Dispatch. Demopolis, AL. June 12, 1897.
Huntsville Telegram. Huntsville, AL. September 3, 1919.
________. September 7, 1919.
Mobile Register. Mobile, AL. November 1907.
________. September 21, 1913.
________. November 6, 1921.
Montgomery Advertiser. Montgomery, AL. March 13, 1886.
________. June 16, 1895.
________. September 15, 1905.
________. June 13, 1906.
________. December 16, 1906.
________. September 22, 1912.
________. March 19, 1913.
________. January 21, 1915.
________. January 22, 1915.
________. January 23, 1915.
________. January 2, 1917.
________. February 24, 1918.
________. July 31, 1919.
________. September 2, 1919.

________. October 10, 1919.
________. February 7, 1920.
________. April 10, 1921.
________. September 4, 1921.
________. April 1, 1923.
________. June 8, 1924.
________. June 11, 1924.
San Francisco Examiner. San Francisco, CA. 1893
State Herald. Birmingham, AL. January 10, 1897.
________. February 21, 1897.
________. March 6, 1897.
Selma Times. Selma, AL. September 15, 1885.
_______. "Green Sand". W. S. Purifoy. September 15, 1885.
Southern Signal. Prattville, AL. December 16, 1881.
Tuscaloosa News. Tuscaloosa, AL. June 13, 1906.
________. January 9, 1911.
________. September 16, 1911.
________. May 31, 1912.
________. September 22, 1912.
________. August 21, 1913.
________. January 1915.
________. April 17, 1918.
________. March 24, 1919.
________. June 8, 1919.
________. September 7, 1919.
________. April 6, 1924.
________. December 15, 1930.
________. June 16, 1975.
________. October 27, 1991.
Tuscaloosa Times – Special Edition. Tuscaloosa, AL. October 28, 1896.
________. Trade Edition. Tuscaloosa, AL. May 12, 1899.
Tuscaloosa Times-Gazette. Tuscaloosa, AL. April 21, 1885
________. June 20, 1896.
________. May 19, 1897.
________. June 7, 1897.
________. February 3, 1902.
________. April 4, 1903.
________. December 13, 1905.
________. December 25, 1905.
________. December 18, 1907.
________. December 25, 1907.
________. January 14, 1908.
________. April 26, 1908.
________. April 28, 1908.
________. July 1, 1909.
________. September 2, 1919.

BOOKS

Adams, Charles. *Blocton: History of a Coal Mining Town.* Cahaba Trace Commission. Brierfield, AL. 2001.

Alabama: A Guide to the Deep South. American Guide Series. Writers' Program of the Work Progress Administration of the State of Alabama. Hastings House, New York. 1941.

Armes, Ethel. *The Story of Coal and Iron in Alabama.* Published under auspices of Chamber of Commerce. Birmingham, AL. 1910.

Booth, Jonna J. *Alabama Cemeteries.* Crane Hill Publishers. Birmingham, AL. 1999.

Brasfield, Elizabeth Baker. *History of Alabama Central Female College.* University of Alabama Press. Tuscaloosa, AL. 1944.

Brewer, Willis. *Alabama: Her History, Resources, War Record, and Public Men from 1540 to 1872.* Barrett and Brown. Montgomery, AL. 1872.

Carboro, Bob S. and Mike Goodson. *Etowah County Images of America.* Arcadia Press. Charleston. SC. 1998.

Dean, Lewis, Editor. *The Papers of Michael Tuomey.* The Reprint Company, Publishers. Spartanburg, SC. 2001.

DuBose, Joel C., ed. *Notable Men of Alabama,* Vol. 1. The Reprint Company, Publishers. Spartanburg, SC. 1976.

Ellison, Rhoda Coleman. *Bibb County, Alabama The First Hundred Years 1818–1918.* University of Alabama Press. Tuscaloosa, AL. 1984.

Flynt, Wayne. *Mine, Mill and Microchip: A Chronicle of Alabama Enterprise.* Windsor Publications. Northridge, CA. 1987.

Garrett, William. *Reminiscences of Public Men in Alabama.* Plantation Publishing Company's Press. Atlanta, GA. 1872.

Gibson, John M. *Physician to the World: Life of General William C. Gorgas.* University of Alabama Press. Tuscaloosa, AL. 1989.

Griffith, Lucille. *History of Alabama 1540 – 1900, As Recorded in Diaries, Letters, and Papers of the Times.* Colonial Press. Northport, AL. 1962.

________. *Alabama: A Documentary History to 1900.* University of Alabama Press. Tuscaloosa, AL. 1972.

Hames, Carl Martin. *Hill Ferguson: His Life and Works.* University of Alabama Press. Tuscaloosa, AL. 1978.

Hamilton, Virginia Van der Veer. *Alabama: A Bicentennial History.* W. W. Norton and Co. New York. 1977.

Hamilton, Virginia Van der Veer. *Alabama: A History.* W. W. Norton and Co. New York. 1984.

Harris. W. Stuart. *Dead Towns of Alabam*a. University of Alabama Press, Tuscaloosa, AL. 1977.

Howard, Robert M. *Reminiscences.* Gilbert Printing Co. Columbus, GA. 1912.

Hubbs, G. Ward. T*uscaloosa: Portrait of an Alabama County, An Illustrated History.* Windsor Publications. Northridge, CA. 1987.

Jackson, Harvey H. III. *Rivers of History: Life on the Coosa, Tallapoosa, Cahaba, and Alabama.* University of Alabama Press. Tuscaloosa, AL. 1995.

Jones, Walter B. *History and Work of Geological Surveys and Industrial Development in Alabama.* Wetumpka Printing Co. Wetumpka, AL. 1935.

Marks, Henry S. *Who Was Who in Alabama.* Strode Publishers. Huntsville, AL, 1972.

McLaurin, Melton A. and Thomason, Michael V. *The Image of Progress Alabama: Photographs,*

1872–1917. University of Alabama Press. Tuscaloosa, AL. 1980.

Mohr, Charles. Plant *Life of Alabama*. 1901. Vol. 6. Government Printing Office. Washington, D. C. October 1901.

Moore, Albert Barton. *History of Alabama and Her People*. Vol. 1. American Historical Society. Chicago, IL. 1927.

Nobles, Larry W. *Old Autauga – Portrait of a Deep South County*. Cahaba Trace Commission. Brierfield, AL. 2000.

O'Dell, Kimberly. *Calhoun County: Images of America*. Arcadia Publishing. Charleston, SC. 1998.

Owen, Marie Bankhead. *The Story of Alabama*. Vol. 3. Lewis Historical Co. New York. 1949.

__________. Vol. 5. Lewis Historical Publishing Co. New York. 1949.

Owen, Thomas M. *History of Alabama and Dictionary of Alabama Biography*. Vol. 1, 1921. S. J. Clarke Publishing Co., Chicago, IL.

__________. Vol. 2. 1921

__________. Vol. 4, 1921.

__________. Vol. 4, The Reprint Co. Spartanburg, SC. 1978.

Porter, Benjamin F. *Reminiscences of Men and Things in Alabama: A Basic Historical Document (1829–1853)*. Edited by Sara Faneuil Walls. Portals Book. 1983.

Rogers, William Warren; Atkins, Leah Rawls,;Ward, Robert David; and Flynt, Wayne. *Alabama: The History of a Deep South State*. University of Alabama Press. Tuscaloosa, AL. 1994.

Sellers, James B. *History of the University of Alabama* Vol. 1. University of Alabama Press, Tuscaloosa, AL.

__________. Vol. 2. Unpublished. Hoole Library. University of Alabama. University, AL.

Smith, Evelyn Futch. *Charn Cuimhne To Our Scots of North Carolina*. E. F. Smith. 1969.

Stewart, John Craig. *Governors of Alabama*. Pelican Publishing Co. Gretna, LA. 1975.

Sulzby, James F. Jr. *Historic Alabama Hotels and Resorts*. University of Alabama Press. Tuscaloosa, AL. 1960.

Wolfe, Suzanne Rau. *University of Alabama: A Pictorial History*. University of Alabama Press. Tuscaloosa, AL. 1983.

Correspondence

Smith, Eugene Allen. Letter Collection in the Geology Library. University of Alabama, Tuscaloosa, AL.

Smith, Eugene Allen. Scrapbooks. Hoole Library. University of Alabama, Tuscaloosa, AL.

Public Documents

Butts, Charles. "Memorial of Eugene Allen Smith." *Bulletin of the Geological Society of America*. 39:51–65. March 30, 1928.

Lloyd, Stewart J. *Eugene Allen Smith Alabama's Great Geologist*. Newcomen Society in North America. New York, 1954.

Smith, Eugene Allen. "Journal of Dr.Eugene Allen Smith." Vol. 1. Unpublished.

_______. "Field Notes of Eugene A. Smith." Vol. 2. 1876–1882. Unpublished.

_______. "Report of Eugene Allen Smith, State Geologist, on the Geologcal and Agricultural Survey of Alabama." 1873. Arthur Bingham, State Printer, Montgomery, AL. David P. Lewis, Governor of Alabama.

_______. "Report of Progress for 1874." Geological Survey of Alabama. W. W. Screws, State Printer. Montgomery, AL. 1875. George S. Houston, Governor of Alabama.

_______. "Report of Progress, 1875."

_______. "Report of Progress, Geological Survey of Alabama, 1876."

_______. "Report of Progress, 1877 and 1878." Geological Survey of Alabama. Montgomery, AL. Garrett and Brown, State Printers. Rufus W. Cobb, Governor of Alabama.

_______. "Report for 1879–1880". Geological Survey of Alabama. 1881. Rufus Cobb, Governor of Alabama.

_______. "Report of Progress 1884-88." Geological Survey of Alabama. Montgomery, AL. W. D. Brown and Company. State Printers and Book Binders. 1888. Thomas Seay, Governor of Alabama.

_______. "Memoir of Daniel W. Langdon, Jr." *Bulletin of the Geological Society of America.* 21:12–16.

_______. "Memoir of Henry McCalley." *Bulletin of the Geological Society of America.* 2:12–16.

Thesis

Mansell, Jeff. A Canebrake Chronicle. "A Brief History of a Region of the Alabama Black Belt." A Thesis presented to the Graduate Faculty of Middle Tennessee State University. May 1997.

Wills

The will of Mary Phelps, Bloomfield, Connecticut. March 31, 1868.

Minutes

Minutes of Board of Trustees' Meeting. University of Alabama. September 1879.

_______. June 1883.

Gazetteer

The following Alabama towns and corresponding counties are mentioned in the text of this book. This is not a comprehensive list. Of course, on his summer expeditions, Eugene Allen Smith visited many more places than are mentioned in the text. Some of the places are now ghost towns. Where the same place name is used in several counties, only the counties visited are listed in this gazetteer.

There have been minor changes to a few counties since Smith's time. The Alabama county map on page 243 depicts the counties as of 2011.

Town	County
Abbeville	Henry
Aberfoil	Bullock
Abner	Tallapoosa
Akron	Hale
Aldrich	Shelby
Alexander City	Tallapoosa
Allenton	Wilcox
Allentown	Mobile
Andalusia	Covington
Anniston	Calhoun
Ararat	Choctaw
Arbacoochee	Cleburne
Arkadelphia	Cullman
Ashland	Clay
Ashville	St. Clair
Athens	Limestone
Auburn	Lee
Barnesville	Tallapoosa
Barton	Colbert
Bay Minette	Baldwin
Bay St. John (Oyster Bay)	Baldwin
Bayou LaBatre	Mobile
Bell's Mills	Cleburne

Benton	Lowndes
Berlin	Dallas
Berry Station	Fayette
Bessemer	Jefferson
Beulah	Lee
Bibbville	Bibb
Big Springs	Randolph
Bigbee	Washington
Birmingham	Jefferson
Bladon Springs	Choctaw
Blake's Ferry	Clay
Blakely	Baldwin
Blockton	Bibb
Blountsville	Blount
Blue Creek	Jefferson
Bluff Springs	Clay
Bluffton	Cherokee
Boiling Springs	Calhoun
Boligee	Green
Bradford	Coosa
Brewton	Escambia
Brierfield	Bibb
Brockton	Coffee
Brookwood	Tuscaloosa
Brownsville	Coosa
Butler	Choctaw
Butler Springs	Butler
Cahaba	Dallas
Calera	Shelby
Caltapa	Pike
Camden	Wilcox
Carbon Hill	Walker
Carthage	Hale
Central	Elmore
Centreville	Bibb
Chamber's Spring	Clay
Chandler's Spring	Talladega
Childersburg	Shelby/Talladega
Choco (Shocco Springs)	Talladega
Choctaw	Clarke
Chulafinnee	Cleburne
Citronelle	Mobile
Clairborne	Monroe
Clairborne Bluff	Monroe

Clanton	Chilton
Clayton	Barbour
Clements Station	Tuscaloosa
Clifton	Wayne
Clintonville	Coffee
Clinton	Greene
Coal Bluff	Wilcox
Coal City	St. Clair
Coaling	Tuscaloosa
Coden	Mobile
Coffee Springs	Geneva
Coffeville	Clarke
Coletas	Clay
Columbiana	Shelby
Coosada, Old	Elmore
Corn House	Randolph
Corona	Walker
Cottondale	Tuscaloosa
Courtland	Colbert
Cuba	Sumter
Cullom Springs	Choctaw
Cussetta	Chambers
Dadeville	Tallapoosa
Daphne	Baldwin
Daviston	Tallapoosa
Dayton	Marengo
Deatsville	Autauga
Delta	Clay
Demopolis	Marengo
Dogwood	Shelby
Dothan (Popular Head)	Houston
Douglasville	Baldwin
Dudleyville	Tallapoosa
Dutch Bend	Autauga
Ecko	Dale
Elba	Coffee
Enitachopco	Clay
Enslen	Walker
Ensley	Jefferson
Enterprise	Coffee/Dale
Epes	Sumter
Equality	Coosa
Eufaula	Barbour
Eutaw	Greene

Evergreen	Conecuh
Fairhope	Baldwin
Fairview	Cullman
Fayette	Fayette
Fayetteville	Fayette
Flat Rock	Jackson
Fleetwood	Tuscaloosa
Flomaton	Escambia
Florala	Covington
Forkland	Greene
Frankfort	Franklin
Franklin	Henry/Macon/Monroe
Frankville	Washington
Fredonia	Chambers
Ft. Gains	Mobile
Ft. Mitchell	Russell
Ft. Payne	DeKalb
Gadsden	Etowah
Gainesville	Sumter
Gainstown	Clarke
Gant	Talladega
Gap	Tallapoosa
Gaston	Sumter
Gate City	Jefferson
Gaylesville	Cherokee
Geneva	Geneva
Glastonburg	Wilcox
Glen Allen	Fayette/Marion
Glendon	Mobile
Glenwood	Crenshaw
Gold Branch	Coosa
Goldville	Tallapoosa
Goodwater	Coosa
Greenpond	Bibb
Greensboro	Hale
Grove Hill	Clarke
Guntersville	Marshall
Gurley	Madison
Hackleburg	Marion
Hackneyville	Tallapoosa
Hamburg	Perry
Hanover	Coosa
Hartford	Geneva
Havana	Hale

Healing Springs	Washington
Hefflin	Cleburne
Helena	Shelby
High Shoals	Randolph
Hissop	Coosa
Hoboken	Marengo
Hollins	Clay
Holt	Tuscaloosa
Huntsville	Madison
Hurtsville	Russell
Intercourse	Sumter
Ironaton	Talladega
Jackson	Clarke
Jacksonville	Calhoun
Jameston	Cherokee
Jasper	Walker
Jemison	Chilton
Johns	Jefferson
Jonesboro	Franklin, Jefferson
Kellerman	Tuscaloosa
Kingston	Jefferson
Kingston, Old	Autauga
Kymulga	Talladega
Lafayette	Chambers
LaGrange	Colbert
Leeds	Jefferson
Leighton	Colbert
Lick Skillet	Madison
Lily Shoals	Bibb
Linden	Marengo
Live Oak	Crenshaw
Livingston	Sumter
Louina	Randolph
Lower Peachtree	Wilcox
Loxley	Baldwin
Luverne	Crenshaw
Mad Indian	Randolph
Magnolia Springs	Baldwin
Malcolm	Washington
Marion	Perry
Marlowe	Baldwin
Maylene	Shelby
Mill Town	Chambers
Milldale	Tuscaloosa

Millville	Sumter
Mineral Springs	Chilton
Mobile	Mobile
Mon Louis	Mobile
Monroeville	Monroe
Monte Sano	Madison
Montevallo	Shelby
Montgomery	Montgomery
Montrose	Baldwin
Morrisville	Calhoun
Mott's Mill	Lee
Moundville	Hale
Mt. Meigs	Montgomery
Mt. Olive	Jefferson
Mt. Vernon	Washington
Munford	Talladaga
Muscle Shoals	Colbert
Naheloa	Choctaw
Needmore	Lawrence
Newbern	Hale
Notasulga	Lee/Macon
Oak	Baldwin
Oak Bowery	Chambers
Old Cahaba	Dallas
Old Coosada	Elmore
Old Erie	Greene
Old Kingston	Autauga
Old Socopatoy	Coosa
Old St. Stephens	Washington
Old Warrior Town	Cullman
Old Washington	Autauga
Opelika	Lee
Opp	Covington
Orange Beach	Baldwin
Orrville	Dallas
Oxford	Calhoun
Oxmoore	Jefferson
Oyster Bay (Bay St John)	Baldwin
Ozark	Dale
Palos	Jefferson
Perdido	Baldwin
Perdue Hill	Monroe
Perote	Bullock
Peterson (Ruhame)	Tuscaloosa

Piedmont	Calhoun/Cherokee
Pine Hill	Wilcox
Pittsview	Russell
Plantersville	Dallas
Pleasant Hill	Dallas
Pocahontas	Walker
Pollard	Escambia
Popular Head (Dothan)	Houston
Prairie Bluff	Wilcox
Prattville	Autauga
Pyriton	Clay
Quenelda	Clay
Ragland	St. Clair
Rescueville	Choctaw
Riverside	St. Clair
Roanoke	Randolph
Roberts	Escambia
Robinson Springs	Elmore
Rock Run	Cherokee
Rose Hill	Covington
Ruhama (Peterson)	Tuscaloosa
Sand Fort	Russell
Sand Tuck (Santuck)	Elmore
Sanders Ferry	Tuscaloosa
Scottsville	Bibb
Searles	Tuscasloosa
Selma	Dallas
Sheffield	Colbert
Shinbone Valley	Cherokee
Shocco Springs (Choco)	Talladega
Siluria	Shelby
Slocomb	Geneva
Snow Hill	Wilcox
Socopatoy	Coosa
Spring Hill	Mobile
Squaw Shoals	Tuscaloosa
St. Stephens, Old	Washington
Statesville	Autauga
Stewart	Hale
Stockton	Baldwin
Suggsville	Clarke
Sumterville	Sumter
Sunny South	Wilcox
Sylacauga	Talladega

Sylvan	Tuscaloosa
Talladega	Talladega
Talladega Springs	Talladega
Tallahatta Springs	Clarke
Tallassee	Elmore/Tallapoosa
Tannehill	Tuscaloosa
Taylorsville	Tuscaloosa
Thomaston	Marengo
Thomasville	Clarke
Tishabe	Greene
Toadvine	Jefferson
Tomkinsville	Choctaw
Troy	Pike
Tuckersburg	Chambers
Tunnel Springs	Monroe
Tuscumbia	Colbert
Tuskegee	Macon
Union	Greene
Union Springs	Bullock
Uniontown	Perry
Vance	Tuscaloosa/Bibb
Vance's Station	Tuscaloosa
Verbena	Chilton
Village Springs	Jefferson
Vine Hill	Autauga
Vinegar Bend	Washington
Vinton	Autauga
Waldo	Talladega
Walnut Hill	Tallapoosa
Warsaw	Sumter
Washington, Old	Autauga
Watson	Monroe
Wedowee	Randolph
Weogufka	Coosa
Wetumpka	Elmore
Whatley	Clarke
White Bluff	Marengo
White Plains	Calhoun
Winchester	Madison
Woods Bluff	Clarke
Woodstock	Bibb
Yolande	Tuscaloosa
York	Sumter

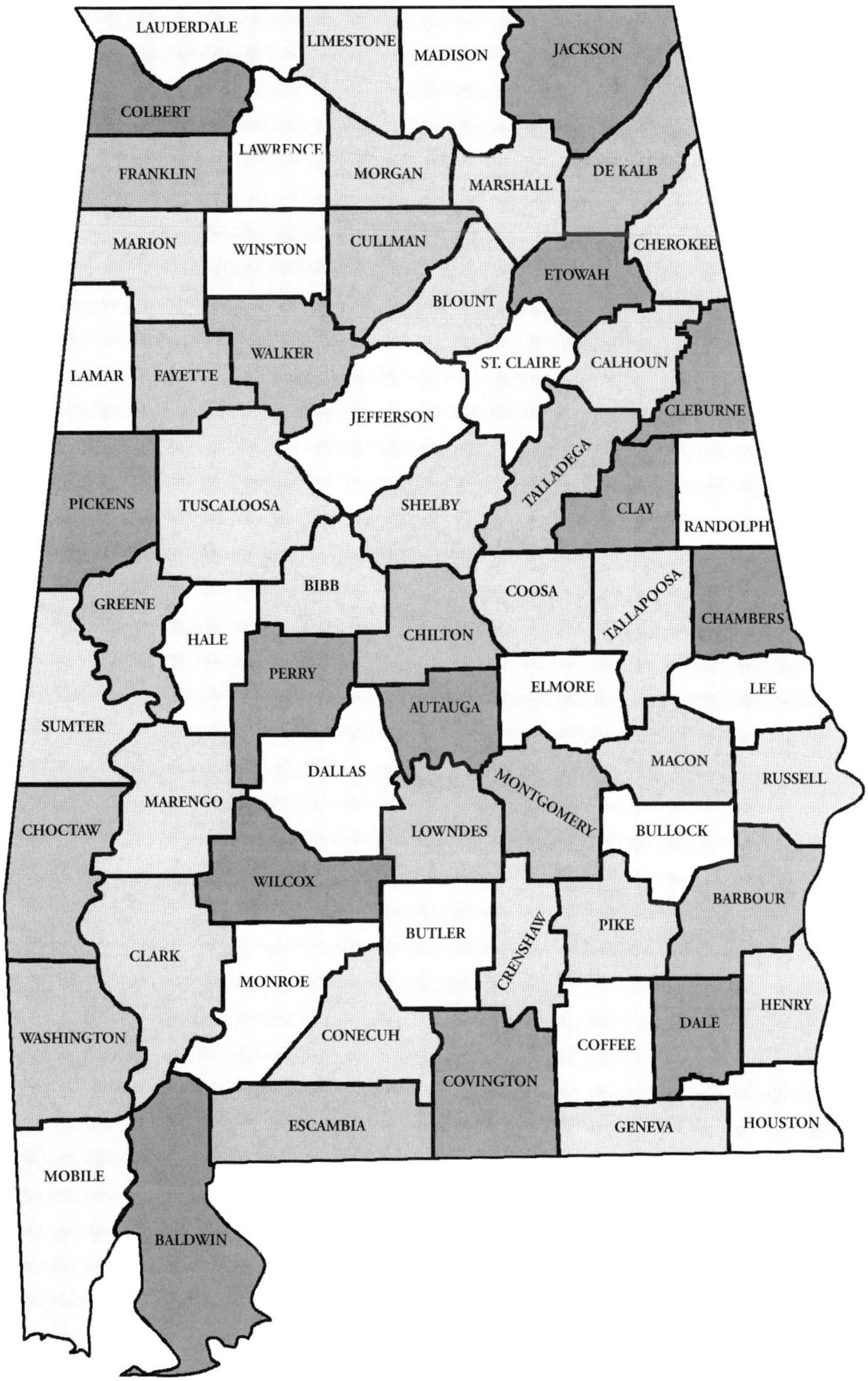
LAUDERDALE
LIMESTONE
MADISON
JACKSON
COLBERT
LAWRENCE
FRANKLIN
MORGAN
MARSHALL
DE KALB
MARION
WINSTON
CULLMAN
CHEROKEE
ETOWAH
BLOUNT
LAMAR
FAYETTE
WALKER
ST. CLAIRE
CALHOUN
JEFFERSON
CLEBURNE
PICKENS
TUSCALOOSA
SHELBY
TALLADEGA
CLAY
RANDOLPH
BIBB
GREENE
COOSA
TALLAPOOSA
CHAMBERS
HALE
CHILTON
PERRY
ELMORE
LEE
AUTAUGA
SUMTER
DALLAS
MACON
RUSSELL
MONTGOMERY
MARENGO
CHOCTAW
LOWNDES
BULLOCK
WILCOX
BARBOUR
BUTLER
PIKE
CRENSHAW
CLARK
MONROE
HENRY
DALE
CONECUH
WASHINGTON
COFFEE
COVINGTON
ESCAMBIA
GENEVA
HOUSTON
MOBILE
BALDWIN

Index

Editors' Note: Place names not including a state name are assumed to be in Alabama. Parenthetical kinship notations are in relation to Eugene Allen Smith Sr. .

M

N

O

P

T

U